World Regional GEOGRAPHY
AN INTRODUCTION

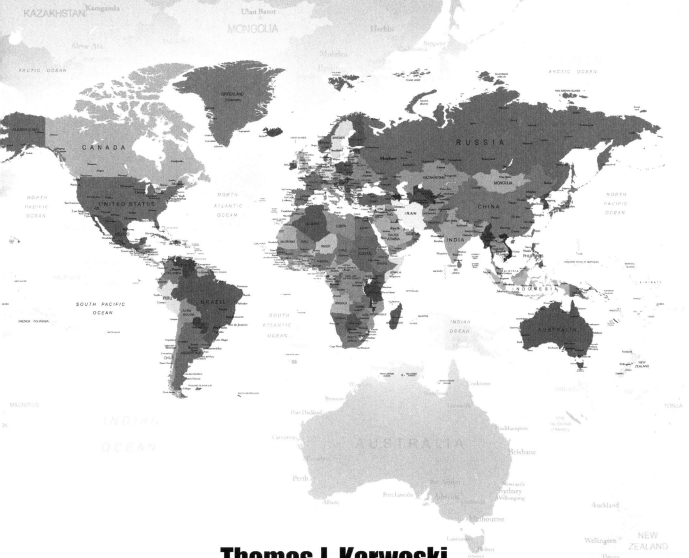

Thomas J. Karwoski
Anne Arundel CC

Kendall Hunt
publishing company

Cover image 2016, Shutterstock.com

www.kendallhunt.com
Send all inquiries to:
4050 Westmark Drive
Dubuque, IA 52004-1840

Printed in the United States of America

Dedication

This book is dedicated to my daughter, Kristin Roemer; my son-in-law, Sean Roemer; and my granddaughters, Kaitlyn Roemer and Hannah Roemer, for their love, support, and encouragement.

Acknowledgments

Sincere thanks to the following students in my World Regional Geography classes in the fall of 2015 for their invaluable feedback, comments, and editing for the first two chapters: Alanna Baker, Jessica Dennison, Justin Hachey, Patricia Hill, Christian Johnson, Elizabeth Kirkpatrick, and Alexis Terry. My apologies to any students I may have inadvertently omitted.

Contents

INTRODUCTION TO GEOGRAPHY

What Is Geography?

Geography is a scientific discipline that attempts to describe and explain the locations, patterns, and interactions of places, people, and environment.

For geographers, the three basic areas of concern are locations, patterns, and relationships. The three main questions for geographers are the three "W's":

▶ **Where?** (locations) "Where is this place located?" "Where do you find a group of people, and where do you find an environmental feature?"

▶ **What?** (patterns) "What kind of pattern occurs in the location of places?" "What is the nature of the distribution of people in an area?" "What relationships exist at a place between people and their environment?" "What relationships exist between places?"

▶ **Why?** (relationships) "Why are places, people, and environmental features located where they are?" "Why do you have this type of specific pattern or relationship?"

What makes geography distinctive is not the subject matter it studies but the approaches (perspectives) through which geographers view the world. The most basic and fundamental perspective in geography is the **locational or spatial perspective.** More so than any other discipline, geography studies the spatial (locational) aspects of places plus their patterns and interactions. The geographer's concern with patterns of physical and human phenomena provides a second distinctive perspective, the **regional perspective.** Finally, the concern with people-environment relationships provides a third distinctive approach, the **ecological perspective.**

Types of Location

Location is the most fundamental or basic concept in geography. Location answers the question "where?" There are two main types of location: absolute and relative. **Absolute location** refers to the specific, precise physical location of a place; Anne Arundel Community College is located at 101 College Parkway, Arnold, MD, or located at 39 degrees North latitude and 76 degrees West longitude. **Relative location** refers to the position of a place relative to other known places; Anne Arundel Community College is located about 5 miles north of Annapolis and about 20 miles south of Baltimore. Often, relative location is a more relevant and meaningful way of measuring location.

What Is a Region?

A **region** is a bounded-off area that consists of similar kinds of things, such as similar climate or topography or language. It is not a phenomenon that naturally exists but it is an intellectual concept, a mental construct. A person recognizes a region based on his (her) subjective view of a place. It is a very important tool in geography for classifying and organizing information about the earth. Regions have locational characteristics and are "somewhere." They are territories with boundaries. The boundaries can be sharply defined (such as the boundary between the Great Plains landform region and the Rocky Mountains landform region—the Front Range of the Rockies). Usually boundaries are transition zones. For example, the boundary that separates the Baltimore urban region from the Washington urban region is a transition zone that cuts through central Anne Arundel County, northern Prince George's County, and central Howard County.

There are two general types of regions: formal (uniform) and functional (nodal). A **formal (uniform) region** refers to a bounded-off area of distinctive properties that are evenly or uniformly arranged. Most physically-based regions, such as climate, landform, or soils regions, are formal as are land-use regions and agricultural regions. A **functional (nodal) region** refers to a bounded-off area with distinctive properties that are concentrated in a focus or node with functional connections to the periphery of the area. Banking regions of the United States and urban regions are very good examples of nodal regions.

Map Scale

Map scale refers to the ratio or proportion between map distance and size and real earth distance and size. Maps can be classified as either small-scale or large-scale. **Small-scale maps** show large areas with less detail. All globes and world maps are small-scale images. **Large-scale maps** show small areas with lots of detail. Topographic maps and neighborhood maps are good examples of large-scale maps.

Theory of Plate Tectonics

Until the 1950s the prevailing earth science theory was of a fixed and rigid earth. Today most earth scientists believe in the **theory of plate tectonics,** one of the most important 20th century earth science theories. The theory states that the earth's **lithosphere** (the crust and upper mantle) is comprised of a set of twelve to fifteen plates that shift or move over or under each other.

Plate tectonics theory incorporates two main components: continental drift and the process of sea floor spreading.

Continental drift refers to a set of ideas highly associated with a German scientist named **Alfred Wegener**. In the early 1900s Wegener proposed that the earth has not always looked the way it does today. Over 200 million years ago, he proposed that all of the land areas of the earth were joined together in a supercontinent known as **Pangaea**. Over tens of millions of years, he believed that this supercontinent eventually began to split apart into segments that moved or drifted. The puzzle-like fit of the continental edges, identical plants and animals in widely separated areas, identical rock types in widely separated areas, and a variety of climatic evidence all seemed to support Wegener's ideas. However, his ideas did not conform to the prevailing theory of the time that the earth was rigid and fixed in nature. Very importantly, he could never identify a force that could cause the drifting of the continents.

By the 1950s and 1960s, **Harry Hess**, a Princeton University geologist, was aware of the great topographic diversity of the oceans: mid-ocean mountain ridges and deep oceanic trenches along the edge of the Pacific Ocean. Based on his research, he proposed a "convection" model. He theorized that hot fluid magma was continuously being pushed upward in the earth's interior by internal forces. As this magma rose, it began to cool and harden. As it reached the mid-ocean ridges, it solidified into new sea floor, causing the pre-existing sea floor to be pushed aside. As the sea floor began to move away from the mid-ocean ridges, it caused the plates to shift or move. This process is now referred to as **sea floor spreading**. Supporting evidence based on ocean floor age and magnetism seems to strongly indicate the existence of this process. Most earth scientists now believe that sea floor spreading is the force that Wegener could not identify.

The theory of plate tectonics is closely associated with other geologic activity such as mountain-building, earthquakes, and volcanoes. Most of the earth's major mountain systems such as the Andes, Alps, and Himalayans are connected to plate activity. A circular-shaped area around the rim of the Pacific Ocean basin is where the world's greatest concentration of volcanic and earthquake activity occurs. This area is known as the **Ring of Fire**. The Ring of Fire results from dense oceanic plates sliding beneath lighter continental plates creating geologic instability such as earthquakes and volcanoes.

The Pleistocene

The greatest degree of world glacial activity occurred during a relatively recent time in earth history known as the **Pleistocene**. The Pleistocene began about 1.5–2 million years ago and ended about 10,000–20,000 years ago. This geologic period is commonly referred to as The Ice Age. During the Pleistocene, most of northern Russia, northern Europe, Canada, and the northern edge of the continental United States were covered by large ice sheets. In North America, the southern-most extent of this glaciation extended to the Missouri and Ohio Rivers.

Climate Controls

Climate is the result of the interaction of a number of factors known as climate controls.

The most basic control on climate is **latitude**. The angle of sun's energy and the length of daylight hours are both mainly affected by latitude. Latitude essentially determines basic temperature conditions. As latitude increases, temperature tends to decrease.

Another climate control is **location relative to water**. **Maritime locations**, places near large bodies of water, have moderate seasonal temperature changes because water heats and cools slowly. Maritime locations include places like San Francisco, Seattle, and Rome, Italy. **Continental locations**, places far from large bodies of water, have great seasonal temperature changes because land heats and cools quickly. Continental locations include places like central Canada, eastern Europe, and most of Russia.

A third climate control is **altitude (elevation)**. Air at lower altitudes is warmer and holds more moisture. Air at higher altitudes is cooler and holds less moisture. In the tropical latitudes places at low altitudes are very warm and very wet. In the tropical latitudes, places at higher altitudes are cooler and drier.

A fourth climate control is **landforms**. If air encounters a landform such as a mountain, it is forced to rise and the air cools, causing precipitation on the **windward side** of the mountain. As the air descends the other side, the **leeward side**, the air warms and dries out. This creates a dry condition known as the **rain shadow effect**. In the western United States places such as Portland and Seattle are lush and green because they are on the windward sides of the Pacific mountains. In contrast eastern Washington and eastern Oregon are dry and brown because they are on the leeward sides of the mountains.

A last climate control is the effect of **prevailing wind and ocean currents**. Warm air and warm ocean currents bring in warm and wet conditions to an area. In contrast, cold air and cold ocean currents bring cold and dry conditions to an area. The southeast United States is hot and wet in the summer because of the warm Gulf Stream ocean current and winds coming from the Gulf of Mexico. The southwest coast of the United States is dry in the summer with lower summer temperatures because of the cold California ocean current and winds coming off the ocean.

Climate Regions

Globally, the world is characterized by four main climate regions. The **humid equatorial (tropical) climate** occurs in the tropics from about the equator to 25 degrees of latitude. It is mainly very warm and very wet. Examples include the Amazon Basin in South America, the Congo Basin in central Africa and the islands of Indonesia. **Dry climates** occur in both tropical and middle latitude areas of the world. They are found in places such as the Sahara in northern Africa, central Australia, southwest North America, southwest South America, and central Asia. They are characterized by little annual precipitation (generally less than 20 inches per year) and there is more evaporation than precipitation. **Humid temperate (mild) climates** are found between 25 degrees of latitude and 50 degrees of latitude. Examples include most of the eastern U.S., southern and western Europe, southeast Australia, and east China. They are climates with moderate levels of temperature and precipitation that create the conditions for most agricultural production. **Humid cold climates** are found in the interiors of landmasses and upper high latitude areas. Examples include Russia, Canada, far northern Europe, most of eastern Europe, and north China. They are also known as **continental climates**, with large seasonal changes in temperature and relatively lower amounts of precipitation.

World Population

There are approximately **7.5 billion people in the world**. A striking feature of this world population is its uneven distribution. Most people are found in middle latitude coastal areas in the Northern Hemisphere. There are three main world population clusters. **South Asia** is the most populous cluster, with about 25 percent of the world's population. It includes the second most populated country, India. **East Asia** is the second largest cluster, with almost 25 percent of the world's population. It includes China, currently the most populated country in the world. The third main cluster is **Europe**, including the European part of Russia. This cluster has about 15 percent of the world's population.

Population change can result from either natural change or migration. **Natural change** is the difference between fertility and mortality in a place. The most common way to measure fertility is through the **crude birth rate** (the # of births/1000 people). The most common way of measuring mortality is through the **crude death rate** (the # of deaths/1000 people). The difference between these two rates is the **rate of natural change** (the crude birth rate – the crude death rate, typically expressed as a percentage). The best way to measure fertility is the total fertility rate. **Total fertility rate** refers to the average number of children a woman has between the ages of 15–44. A total fertility rate of "2.1" is known as the **replacement rate**. Over time, this rate would likely lead to population stability.

Another very important way of measuring mortality is the infant mortality rate. **Infant mortality rate** refers to the number of infants/1000 births who die before age 1.

Population structure includes features such as the **age structure** (the proportion of a population under age 15 and over 65); the **sex structure** (the proportion of males/100 females); and **life expectancy** (the average number of years a person can expect to live).

Demographic Transition Model

In the mid-twentieth century, a population change model was introduced—the **demographic transition model**. It was based on the actual demographic experience of European and European-related countries that have a long and accurate set of census data. The demographic transition model links population change to economic change over time. Today, five stages are commonly identified for the model.

Stage 1 includes countries that are dominantly agricultural, rural, and isolated. Given these conditions, stage 1 countries have high birth rates and equally high death rates. The result is population stability but at a high level. Only a few isolated and very poor countries are today in stage 1.

Stage 2 is the first major transition that occurs. Countries are still dominantly agricultural and rural so birth rates remain high. However, countries are no longer isolated and medical technology is introduced allowing a quick and sharp decline in death rates. With high birth rates and fast falling death rates, the population grows very fast. Stage 2 is a stage of **population explosion**. Most of the less developed countries, like countries in Africa, Latin America, and much of Asia, are in stage 2.

Stage 3 is the second major transition that occurs. Death rates continue to fall and birth rates begin to decline. The birth rate decline results from a country undergoing a transition to industrialization and urbanization. As European and European-like countries shifted from agriculture to industry and from rural to urban life, birth rates fell. The population still grows, but at a slower rate. Countries like Brazil and China are in stage 3.

Stage 4 includes countries that now are highly economically-developed and dominantly urban and suburban. Birth rates are low and death rates are low, creating population stability but at a low level. Most of the more developed countries are in stage 4.

Stage 5 includes a number of more developed countries whose birth rates are so low combined with large aging populations that the birth rate is now below the death rate, leading to population decline. Stage 5 is a stage of **population implosion**. Many countries in Europe, along with Japan and Russia, are in an implosion stage.

World Languages

Language is most commonly classified on the basis of its genesis or origin. The broadest level of language classification is a language family. A **language family** is a relatively large and broad assortment of all languages related by common ancestry. The most spoken and most widespread language family is the **Indo-European language family**. It dominates in Europe, the Americas, the most populated parts of Russia, as well as in Iran, Afghanistan, and most of south Asia. The **Afro-Asiatic language family** dominates in northern Africa and much of southwest Asia. The **Niger-Congo language family** dominates in Africa below the Sahara. The **Uralic language family** dominates in Finland, Estonia, and Hungary. The **Altaic language family** dominates in Turkey and Central Asia. The **Dravidian language family** dominates in southern India. The **Khoisan language family** dominates in southwest Africa. The **Sino-Tibetan language family** dominates in China and parts of mainland Southeast Asia.

World Religions

Scholars commonly identify five main world religions. They are Christianity, Islam, Hinduism, Buddhism, and Judaism. **Christianity** is the largest and most widespread religion. It dominates in Europe, the Americas, most of the populated parts of Russia, as well as Africa below the Sahara. **Islam** is the world's second largest and second most widespread religion. It dominates in northern Africa, southwest Asia, central Asia, parts of south Asia, and Indonesia. **Hinduism** is the world's third largest religion in numbers and dominates in India and Nepal. **Buddhism** is the world's fourth largest religion in numbers and dominates in East Asia and mainland Southeast Asia. **Judaism** is relatively insignificant in numbers and only dominates in Israel.

Religion is classified in several different ways. On the basis of location origin and dominance, religion is differentiated into eastern and western religions. **Eastern religions** have their origin and dominance in southern and/or eastern Asia. Hinduism and Buddhism are the two main world eastern religions. **Western religions** have their origin in southwest Asia. Judaism, Christianity, and Islam are the three main world western religions. On the basis of philosophy, religion is differentiated into monotheistic and polytheistic religions. **Monotheistic religions** believe in one god. Judaism, Christianity, and Islam are the three main world monotheistic religions. **Polytheistic religions** believe in many gods or spirits. Hinduism is the only main world polytheistic religion. Buddhism is not clearly classified on the basis of philosophical belief. A third type of religion classification is based on whether the religion is localized to a certain group or area without conversion or missionary activity or whether a religion is universally applicable with much conversion or missionary activity. An **ethnic religion** is localized without intense conversion activity. Hinduism and Judaism are two main world ethnic religions. **Universalizing religions** are universally applicable to anyone anywhere and are very conversion-oriented. Christianity, Islam, and Buddhism are the three main world universalizing religions.

Forms of Territoriality

Territoriality is the process by which an individual or group identifies with an area, claims it as their own and defends it if needed. Over time, several forms of territoriality have emerged. **State** is an inherently political concept, comprised of an area, recognized boundaries, and a government in effective control of the area. Essentially, a state is a country. **Nation** is a social-cultural concept, comprised of a group of people bonded or unified by one or more items. Beginning with the French Revolution in Europe, a new form of territoriality emerged known as a nation-state. A **nation-state** is a country comprised of a unified group of people. There are different levels of nation-states. A **unified nation-state** refers to a country where almost everyone is ethnically and culturally the same. Japan, Denmark, and Slovenia are examples of unified nation-states. The United States is broadly a unified nation-state since most Americans are unified by the English language, Christian religion, and similar values, such as democracy, liberty, and freedom. A **binational state** is a country comprised of two distinct nations. Canada, Belgium, and Croatia are examples. A **multinational state** is a country comprised of three or more distinct nations. Bosnia, Serbia, and every country in Africa are examples. A **stateless nation** occurs where a nation does not have its own state. Examples include the Palestinians, the Kurds, and the Basques. Especially since the end of World War II, a new form of territoriality exists: a supranational organization. A **supranational organization** is a collection of three or more countries (states) that join together for some mutual economic, military, or political purpose. The United Nations, NATO (North Atlantic Treaty Organization), the European Union, and NAFTA (North American Free Trade Agreement) are good examples.

Globalization

Globalization is a process that involves the global connections of people and places. Key characteristics include transnational corporations and global interaction and diffusion. It is a process that emphasizes free trade and involves much cultural standardization, such as the rapid spread of McDonald's and Coca-Cola across the world. It is a very controversial process that many people claim makes the rich richer and the poor poorer as well as creating more standardization or homogenization of our cultures.

Globalization is all around us. We see its influence in clothing, shoes, cars, cell phones, and all kinds of merchandise. The world today is highly interconnected. What happens in one place has repercussions on much of the rest of the world. If China's economy stalls, the whole world is economically affected.

Sectors of the Economy

Typically, the economy is comprised of three different sectors or activities. **Primary activities** involve the extraction of raw materials directly from the environment. Agriculture, mining, fishing, and logging are primary activities.

Secondary activities include the processing of raw materials into finished products. Manufacturing is identical to secondary activity. Steel production, automobile production, processing of crude oil, and milling of grain into cereal products are all examples of secondary activity.

Tertiary activities include the marketing of products such as retailing and wholesaling. Service activities like medicine, law, and education are also tertiary activities.

Economic Development

Economic development is a process by which countries experience economic change over time. There is a wide range or spectrum of development that exists in the world today. To better understand this range of development, researchers often divide it into two main groups—the more developed countries and the less developed countries.

More developed countries include the United States, Canada, Australia, New Zealand, Japan, South Korea, Taiwan, Singapore, Russia, and the countries of Europe.

Less developed countries include the countries of Middle and South America, the countries of Africa, and most of the countries of Asia.

The position of a country along the spectrum of development is identified by the use of various indicators. **Social indicators of development** include literacy and level of education. More developed countries tend to have higher rates of literacy and higher levels of formal education. Less developed countries have an opposite set of social factors. **Demographic indicators of development** include birth rate, rate of change, life expectancy, age structure, infant mortality, and total fertility rate. The one ineffective demographic factor is the death rate. More developed countries tend to have low birth rates, low or negative population change, long life expectancy, few youth and many elderly people, low infant mortality, and low total fertility. Less developed countries tend to have the opposite set of factors. **Economic indicators of development** include per capita income, proportion of a population that lives in urban areas, and dominant economic sector. More developed countries tend to have high income levels, a high proportion of urban population, and dominance of services and industry. Less developed countries tend to have the opposite set of factors.

An alternate measure of development is through the UN-sponsored **Human Development Index (HDI)**. This is an attempt to better measure the quality of life of an area. It is based on variables such as literacy, average years of schooling, life expectancy, and gross national income per capita. The UN releases its index report (scaled on a magnitude from 0 to 1) annually. Typically, countries with high HDI scores tend to be the more developed countries. Countries with low HDI scores tend to be the less developed countries.

Region Framework

In this course, we will take a **regional approach** to geography. We will examine a number of world regions, both more developed and less developed. **More developed regions** include Europe, Russia, U.S. and Canada, and Australia and New Zealand. **Less developed regions** include Middle America, South America, Sub-Saharan Africa, North Africa and Southwest Asia, South Asia, East Asia, Southeast Asia, and Oceania. For each world region, we will examine important topical features such as climate, landforms, language, religion, politics, and economics. We start our investigation into world regions by examining the more developed region of Europe in Chapter 2.

EUROPE

Where Is Europe?

As a continent, Europe extends eastward from the Atlantic Ocean to the Ural Mountains in west central Russia. It is located north of the Mediterranean Sea, Black Sea, Caucasus Mountains, and Caspian Sea.

As a region, Europe extends eastward from the Atlantic Ocean to the western border of Russia.

Major Qualities of Europe

Europe is a highly-developed world region. It is densely populated with numerous political states. Some states are large in area and population such as Germany, France, and Italy. Other states are small in area and population such as Luxembourg, Lichtenstein, Monaco, and Vatican City. Europe currently has about 600 million people who are highly urbanized with a largely aging, declining population. Historically, it has been a major center of innovation with many significant developments (revolutions) originating in Europe and then spreading throughout the rest of the world.

Climate Patterns

The Europe region is mainly located in the middle latitude area of the world (30 degrees to 60 degrees North latitude).

Most of western Europe has a **humid temperate (mild) marine climate**. Temperatures are moderate through the year with 30–80 inches of annual precipitation. These mild and wet conditions are the result of the moderating effect of adjacent water bodies as well as the North Atlantic Current (Drift). The **North Atlantic Current (Drift)**, a warm ocean current originating in the tropics and moving northeast across the Atlantic Ocean, brings warm and wet weather to most of western Europe. London, Dublin, and Paris are examples.

Most of southern Europe has a **humid temperate (mild) Mediterranean climate**. Temperatures are mild through the year with 15–30 inches of annual precipitation, with a summer drought seasonal pattern. Rome and Athens are examples.

Most of eastern and far northern Europe has a **humid, cold climate**. There are significant seasonal temperature changes with short, hot summers and long, cold winters. Precipitation averages about 20–40 inches annually. These climates are also known as **continental climates** due to their position away from large bodies of water. Stockholm, Helsinki, and Warsaw are examples.

Landform Patterns

The **Western Uplands** largely comprise the western edge of Europe. These are hilly and highland areas such as most of Scandinavia, western parts of the British Isles, Brittany peninsula in France, and the western Iberian Peninsula (Spain and Portugal).

The **Alpine System** comprises most of southern Europe. These are mountainous areas including the Pyrenees along the Spain/France border, Alps in central Europe, Appennines in central Italy, Carpathians in eastern Europe, and the Balkan Peninsula.

The **Central Uplands** in central Europe are located to the north of the Alpine System. These are transitional, hilly areas at the northern edge of the Alpine System and include eastern France and central Germany.

The **North European Lowland (Plain)** is the most significant landform region. This is the main lowland area of Europe, extending from western France through southern England, Belgium, Netherlands, Denmark, northern Germany, Poland, Ukraine, and into European Russia. It is a major historical route of movement from west to east. The North European Lowland includes most of Europe's population, large cities, agriculture, and industry.

European Revolutions

Historically, Europe has been a major hearth of global innovations. The **medieval agricultural revolution** began around 1200–1400. Contour plowing, hybrid strains of grain, and land consolidation were hallmarks of this phenomenon. As people began developing technology and applying new scientific methods it revolutionized farming and greatly improved crop yields.

A **commercial revolution** occurred from 1400–1700. This was based on maritime shipping and trade and coincided with the age of European discovery and exploration.

The very important **industrial revolution** started in northern England in the mid-1700s. It spread to mainland Europe and then was globally diffused through migration of northern and western Europeans. In many ways, it is the uneven spread (diffusion) of the industrial revolution that has resulted in the current global range of more developed and less developed countries. The industrial revolution involved the use of machines in factories; these machines allowed a more effective, efficient, and cheaper way to produce mass volumes of goods.

The **modern urban revolution** is very closely related to the industrial revolution. As Europe began to industrialize, it also began to greatly urbanize. Factories mainly occurred in cities; this event allowed larger proportions of people to live and work in the city rather than in the country. **"Millionaire" cities**, each with at least one million people, began to emerge and spread. By 1800, London became the first millionaire city.

A series of **political revolutions** emerged starting with the French Revolution in the late 1700s. The phenomenon of nationalism began in Europe. **Nationalism** refers to a process where a nation seeks political representation. A new form of territoriality, the **nation-state**, began to appear. Democracy began to be a more common phenomenon in Europe at this time, though its roots are in Greece about 2,000 years ago.

Primate City

A **primate city** refers to a city that is the economic core of a country, has a population more than twice as large as any other city and is often the political capital. London, Paris, and Athens are examples of primate cities in Europe.

Language Patterns

Within the Europe region there is a broad unity of language. The **Indo-European language family** predominates over most of the region. However, within this broad unity there is much regional diversity.

The **Germanic language group** dominates in northern and western Europe. Norwegian, Swedish, Danish, German, Dutch, Flemish, and English are examples of Germanic languages.

The **Romance language group** dominates in southern Europe. Portuguese, Spanish, French, and Italian are examples of Romance languages. In eastern Europe the Romanian language is also a Romance language.

The **Slavic language group** dominates in eastern Europe. Polish, Czech, Slovakian, Slovenian, Croatian, Serbian, and Ukrainian are examples of Slavic languages.

The **Celtic language group** dominates in the remote western edges of the Europe region. Scottish and Irish Gaelic, Welsh, and Breton are example of Celtic languages.

The only **non-Indo-European languages** in Europe are Finnish, Estonian, Hungarian (Uralic language family), and the Basque language family.

Religion Patterns

Similar to language, there is a broad religious unity in Europe but also regional diversity. **Christianity** predominates over most of the Europe region.

Protestant Christianity predominates in northern and western Europe. **Lutheran Protestantism** is dominant in Scandinavia and northern Germany. **Anglican Protestantism** prevails in the United Kingdom.

Roman Catholic Christianity predominates in southern Europe and the northern and western Slavic areas. Portugal, Spain, France, Italy, Poland, Czech Republic, Slovenia, and Croatia are examples.

Orthodox Christianity predominates in the southern and interior parts of eastern Europe. Serbia, Greece, Bulgaria, Ukraine, and Russia are examples.

Islam predominates in Albania and Kosovo.

Population

There are about 600 million people living in the Europe region. Most of the population is found within the **North European Lowland (Plain)**. Given its high level of economic development, Europe is experiencing slow growth or negative growth. Most countries are below replacement rate fertility levels. Thus, much of Europe is in the population implosion stage (stage 5) of the demographic transition model, with very low birth rates and low but rising death rates. For many European countries deaths are annually higher than births.

Political Forces

At any one time, a place is affected by two sets of opposite forces: centripetal forces and centrifugal forces. **Centripetal forces** are any factors that lead to unity, integration, and bonding. Common language, common religion, and common ethnicity are examples of such forces. **Centrifugal forces** are any factors that lead to disunity or division. Different languages, different religions, and different ethnic groups are examples of centrifugal forces.

Supranationalism

An important type of centripetal force is supranationalism. **Supranationalism** is a venture that involves three or more states joining together for some mutual benefit.

The first major example of supranationalism in Europe occurred near the end of World War II. In 1944, three small states in Western Europe, the Benelux countries, formed the Benelux Union. The **Benelux countries** are Belgium, Netherlands, and Luxembourg. The **Benelux Union** was a customs union where the three countries decided to economically unite for economic purposes but remained politically independent.

In 1951 the three Benelux countries joined France, Italy, and West Germany to form a limited economic organization. The **European Coal and Steel Community (ECSC)** was formed to economically unite the coal and steel resources of the six member states. This limited organization worked well enough that the same six states met in Rome in 1957 and signed the Treaty of Rome. The result was the **European Economic Community (EEC)**, also known as the Common Market. By the mid-1960s, the EEC created a council, a court, a parliament, a commission, and changed its name to the **European Community (EC)**.

In 1991, the EC became the **European Union (EU)**. The EU has a common currency (the euro), common foreign policies, a common passport system with free movement between member states, extensive economic integration, and tough environmental laws.

Over time, the European Union has expanded to a twenty-eight-member organization. In the early 1970s, Denmark, the United Kingdom, and the Republic of Ireland joined. In the 1980s, Spain, Portugal, and Greece were admitted, making it a twelve-member group. In the mid-1990s, Austria, Sweden, and Finland joined. In 2004, ten countries joined all at once. These ten new countries were Estonia, Latvia, Lithuania, Poland, the Czech Republic, Slovakia, Hungary, Slovenia, Malta, and the Greek-controlled part of Cyprus. In 2007, Romania and Bulgaria were added. In 2013, Croatia became the twenty-eighth member state. It is easier to remember what parts of Europe are not in the EU. Iceland, Norway, Switzerland, Serbia, and several other states in southeast Europe are not yet members.

Devolution

An important type of centrifugal force is devolution. **Devolution** is a process whereby a large area is divided into smaller areas or whereby a central government provides certain areas with some degree of local autonomy, such as local parliaments. The division of former Czechoslovakia into the Czech Republic and Slovakia and the division of former Yugoslavia into Slovenia, Croatia, Serbia, Bosnia, Macedonia, Montenegro, and Kosovo are examples of dividing large areas into smaller areas. An example of the other type of devolution would be the United Kingdom allowing Scotland, Wales, and Northern Ireland to have their own local parliaments. Another example is Spain, which has allowed the Basque country in the north central part of Spain, Galicia in the northwest, and Catalonia in the northeast to have local parliaments or some local control.

Subregions of Europe

We will examine five different subregions of the Europe region. These smaller areas include Western Europe, the British Isles, Mediterranean Europe, Northern Europe, and Eastern Europe.

Western Europe

Western Europe includes the countries of Germany, France, Netherlands, and Belgium. This subregion is the economic and demographic core of the Europe region. It is dominated by Protestant Christian and Germanic-speaking groups of people.

Germany is the most populated and most economically-developed country in Europe. Within west central Germany is the Ruhr. The **Ruhr** is the economic and demographic core of Germany. It includes a series of large cities closely connected together with steel, metal manufacturing, and chemical processing.

France is the country with the largest land area in Western Europe. It is the most important agricultural and food producing country in Western Europe. Paris is an excellent example of a primate city. Unlike most of Western Europe, France is dominantly Romance-speaking and Catholic Christian.

Netherlands is a small, densely populated country west of Germany. It is one of the Benelux countries. Within the Netherlands there is an important area known as the Randstad. The **Randstad** is the area comprised of the three largest Dutch cities: Amsterdam, Rotterdam, and The Hague. The Randstad is the economic and population core of the Netherlands. **Rotterdam**, at the mouth of the Rhine River, is Europe's largest port and one of the largest ports in the world.

Belgium is another Benelux country, located west of Germany and north of France. Belgium is an excellent example of a binational state. Most people in the north are Flemish, Germanic-speaking, and Protestant Christian. Most people in the south are Walloons, French-speaking, and Catholic Christian. Brussels is the capital city of Belgium and a major headquarter city of the European Union.

British Isles

The **British Isles** includes two major islands, Britain and Ireland. There are two countries that comprise this subregion: the United Kingdom of Great Britain and Northern Ireland and the Republic of Ireland. The Industrial Revolution began in northern England.

The **United Kingdom** (**UK**) includes the territories of England, Scotland, Wales, and Northern Ireland. London is the capital of the UK and is an excellent example of a primate city. During the term of former Prime Minister Tony Blair, Scotland, Wales, and Northern Ireland were allowed to form their own local parliaments. There is still much intense nationalism within Scotland, which recently voted to stay within the UK rather than become independent. The UK is predominantly Anglican Protestant Christian and Germanic-speaking.

The **Republic of Ireland** consists of the southern three-fourths of the island of Ireland and it became independent in the early twentieth century. It is predominantly Catholic Christian and Germanic-speaking, but the Irish Gaelic language is also important in literature and history. During the 1980s and 1990s, the Republic of Ireland experienced remarkable economic growth, quickly shifting from a less-developed economy to a more-developed economy. This quick transformation earned it the nickname, "**The Celtic Tiger**" (reminiscent of South Korea, Taiwan, Hong Kong, and Singapore quickly transforming into the "Asian Tigers"). High-tech manufacturing and research along with finance and tourism fueled the economic transition in Ireland.

Mediterranean Europe

Mediterranean Europe includes the countries of Italy, Spain, Greece, and the island of Cyprus. Romance languages and Catholic Christianity predominate over most of the subregion. This is the part of Europe characterized with a humid temperate (mild) Mediterranean climate. Topographically, it is a hilly and mountainous area.

Italy is the most populated country within Mediterranean Europe. It is also the most economically-developed country in the subregion and the country most connected to the European core. Within Italy, the **Ancona Line** separates the more developed north from the less developed south. It extends from the city of Ancona on the east coast to just south of Rome on the west coast. In the more developed north, the main economic and population core is the **Po River Valley,** including the city of Milan. **Milan** is Italy's main industrial and financial center.

Spain is the larger country that comprises the Iberian Peninsula. It is an excellent example of devolution. The central government in Madrid has allowed Catalonia, the Basque country, and Galicia to have local parliaments. Catalonia, the province in the northeast that includes Barcelona, has been especially active in seeking possible independence.

Greece is the major country on the Balkan Peninsula. The Greek language is Indo-European but not part of the Romance language group. Greeks are also dominantly Greek Orthodox Christian in religion rather than Catholic Christian. Greece has been experiencing lots of economic problems recently, requiring massive financial aid from the rest of the European Union.

The island of **Cyprus** is located in the eastern Mediterranean Sea. It is a good example of a binational territory, with Turkish Cypriots controlling the north and Greek Cypriots controlling the south. In 2004, the Greek part of Cyprus was admitted into the European Union.

Northern Europe

Northern Europe includes the countries of Norway, Sweden, Denmark, and Finland. It is an area commonly referred to as Scandinavia. Germanic languages and Lutheran Protestant Christianity predominate here. Northern Europe has very high standards of living and very high quality of life indicators. Politically it is dominated by democratic socialism with the central governments providing "cradle-to-grave" health insurance and a strong government presence in people's lives. The subregion emphasizes very progressive attitudes, with relatively high proportions of females in government and economic activities. Though dominated by Lutheranism, most of the countries are highly secular (religion being a small part of most people's lives).

Eastern Europe

Eastern Europe includes the countries of Poland, the Czech Republic, Slovakia, Hungary, Romania, Bulgaria, Slovenia, Croatia, Bosnia, Serbia, Montenegro, Macedonia, Kosovo, Albania, and Ukraine. It is dominated by Slavic languages and Christianity. The northern and western Slavs are dominantly Catholic Christian whereas the southern and eastern Slavs are dominantly Orthodox Christian. This subregion was formerly under the control of the former Soviet Union, with a Communist economy and government. Since the early 1990s, it has been transitioning from a Communist command economy and Communist government to a market economy and democracy.

The devolution of former Czechoslovakia into the Czech Republic and Slovakia occurred so smoothly that it is often referred to as the **Velvet Divorce**.

Eastern Europe is an excellent example of a shatter belt. A **shatter belt** is a term that describes an area that has been externally and internally splintered and fractured. The external shattering here occurs largely as a result of the relative location of Eastern Europe. Historically, Germans from the west, Russians/Soviets from the east, and Ottoman Turks from the south have created chronic fracturing and splintering in this subregion. Internally, shattering is a result of a tremendous diversity of different ethnic groups, religions, and languages.

Ukraine has the largest land area within the Europe region. It is strategically located between Russians and Soviets to the east and more democratic states to the west. Internally, western Ukraine is dominantly rural, agricultural, Catholic Christian, and ethnically Ukrainian. Eastern Ukraine is more urban, industrial, and Orthodox Christian with sizable concentrations of ethnic Russians. Recently, Russia has annexed the Crimea peninsula and is supporting ethnic Russian efforts in eastern Ukraine to possibly achieve independence from the Ukraine government. This contributes to the country's political instability.

Ethnic Cleansing

Ethnocentrism is the common tendency of an ethnic group to consider that group to be more important than other groups.

Ethnic cleansing is an extreme form of ethnocentrism in which an ethnic group seeks to remove or obliterate another group to ensure its superiority. Another name for ethnic cleansing is **genocide**.

In Eastern Europe there are two examples of the process of ethnic cleansing—Bosnia and Kosovo.

In **Bosnia,** there are three main ethnic groups: Bosnian Croats, Bosnian Serbs, and Bosnian Muslims (Bosniaks). Up until the early 1990s, these three groups essentially managed to coexist without major problems. With the dissolution of former Yugoslavia in the early 1990s, former republics, such as Bosnia, sought political independence. The Bosnian Serbs, aided by Serbian officials and military, sought to assemble all the Serbs together to form a Greater Serbia. To accomplish this action, the Serbs began to cleanse the area of Croats and especially Muslims. Eventually NATO and UN troops were dispatched to the area to end the Serb ethnic cleansing. The Dayton, Ohio Peace Accords were implemented to divide Bosnia into two main areas of control: a Serb-controlled area and a joint Croat-Muslim area.

In **Kosovo**, the majority of the population was ethnic Kosovars (Albanian-speaking Muslims). Kosovo used to be a political part of Serbia, symbolically representing the core area of early Serb culture. To provide long-term Serbian control of this culturally important area, the Serbs began to cleanse the area of Kosovars. Again, NATO and UN troops were dispatched to end the ethnic cleansing. Eventually, the Kosovars unilaterally seceded from Serbia and formed a politically independent Kosovo.

RUSSIA

Where Is Russia?

As a region, Russia includes the country of **Russia** along with **Transcaucasia** (the three small independent states of Georgia, Armenia, and Azerbaijan). Most of our focus will be on the country of Russia.

Major Qualities of Russia

As a country, Russia is the largest state in size of land area in the world. It has an immense physical size overlapping the continents of Europe and Asia. Russia is the northernmost populous country in the world. Although the country has a very large land area, it has comparatively few people (approximately 140 million). An advantage of its massive land area is a rich variety of resources, including much oil and natural gas. Russia is, in reality, a multiethnic, multicultural state with the Russian ethnic group being the largest in numbers and proportion. Since the early 1900s, Russia has experienced very rapid political, economic, and social change, with most of the change happening since the 1990s.

Landform Patterns

The **Ural Mountains** are a north-south trending mountain system in west central Russia. They are commonly used as a boundary between the continent of Europe to the west and the continent of Asia to the east. Within Russia, they divide European Russia from Asiatic Russia. The Asiatic part of Russia is broadly referred to as **Siberia**. Most of the land area and resources of Russia are in Siberia but most of the people are in European Russia.

The **Caucasus Mountains** are a west-east trending mountain system that lies between the Black Sea and the Caspian Sea. Like the Urals, they are also a boundary between Europe, to the north, and Asia, to the south. They also mark the historical and cultural southern boundary of Russia. Russia is in the north; non-Russia is in the south.

Most of European Russia is comprised of the **Russian Plain**, the eastern extension of the North European Plain. Most of Russia's people, agriculture, and industry are in the Russian Plain.

The **Black Sea** and the **Caspian Sea** are also historic boundaries between the continents of Europe and Asia. The Caspian Sea is the largest lake in the world based on surface area and average volume of water.

Lake Baikal, in South Central Siberia, is the world's deepest lake.

The **Volga River** is the major river of movement of goods and people in Russia. The river flows north to south across the Russian Plain, emptying into the Caspian Sea.

Climate Patterns

Given the latitude and position of most of Russia in the high middle latitudes, the dominant climate type in the Russia region is a **humid, cold climate**, also known as a "**continental**" climate. Summers tend to be short and cool, and winters are long and cold. Precipitation tends to be relatively dry. The land-based nature of most of Russia is a major factor in creating these climate conditions.

Forward Capital

A **forward capital** is a capital city located or relocated away from the historic core toward the periphery, indicating the intent of a state to develop the newer area. In Russia, **St. Petersburg** is a classic example of a forward capital.

Czar Peter the Great decided to locate the capital of the Russian Empire at St. Petersburg, named in his honor, so that it would be Russia's "window to the west." Peter the Great intended to reorient Russia toward Europe and the western side of Eurasia. He brought in architects and designers to create a more European-like city.

Recent Political and Economic Changes

Up until the early 1900s, the Russian Empire was a feudal, medieval state. Czars (tsars) ruled the country with most of the people being poor and landless.

In 1917, in St. Petersburg, the Russian Empire collapsed with the overthrow and assassination of Czar Nicholas and his family. A group of **Bolsheviks** (a faction of Russian Communists) led by Vladimir Lenin, and representing a broad coalition of business people, workers, and peasants, replaced the tsars.

It was **Lenin** who created the **Soviet Union** (the Union of Soviet Socialist Republics) and introduced **communism** as a political and economic system.

A hallmark feature of communism as an economic system was the command economy. The **command economy** was based on centralized economic planning with bureaucrats in Moscow (the new capital) making all economic decisions. Under the command economy, everyone had a job and the government subsidized housing and food. Although it was an equitable system it was very inefficient.

By the 1970s and 1980s economic and social problems increased in scope and numbers. Many people became frustrated with the lack of food, the quality of goods, and the lack of basic liberties such as freedom to travel and freedom of speech. In the late 1980s, **Mikhail Gorbachev** was the Soviet leader. He recognized the numerous problems and believed some action was needed to prevent the collapse of communism. Gorbachev proposed two major strategies to try to save the Soviet Union: glasnost and perestroika.

One strategy was the policy of **glasnost**. Glasnost referred to Gorbachev's intent to open up and liberalize Soviet society. Soviet citizens would have more freedom to travel within and outside the country. More freedom of speech would be encouraged. Greater access to Western goods and clothing would become available.

The other strategy was **perestroika**. Perestroika referred to Gorbachev's intent to restructure the Soviet economy away from a command economy toward a more market-based Western economy.

Before Gorbachev could implement these changes he was replaced by **Boris Yeltsin**, a popular mayor of Moscow. It was actually Yeltsin who began to implement these changes.

In the early 1990s, the Soviet Union collapsed and communism as an economic and political system began to fall. Price controls were removed. State-owned businesses were sold to private investors. People were able to own property and sell goods on their own.

However, many problems and issues have remained. Food is now much more expensive and housing is more expensive. Inefficient factories and businesses have shut down, greatly increasing unemployment. The Russian "Mafia" rapidly increased its presence and influence.

Population

There are about 160 million people in the Russia region, with about 140 million people in the country of Russia and about 20 million people in Transcaucasia. The country of Russia is experiencing population implosion due to low birth rates and rising death rates (especially among middle-age males). Population in Russia will likely continue to fall over the next few decades.

Russian women are having few babies, with a below-replacement fertility rate. Death rates have risen due to exposure to unchecked environmental pollution, increased cancer and suicide rates among middle-age males, and a rapid increase in murders.

Most of the population resides in the European part of Russia, with the attraction of the Russian Plain and more favorable climate and soil conditions there.

Geopolitics

Geopolitics is the study of how place or location affects political behavior. We will examine two geopolitical theories: the heartland theory and the rimland theory.

In the early 1900s, a British geographer, **Halford Mackinder**, introduced the most well-known theory of geopolitics: **the heartland theory**. According to Mackinder, who rules Eastern Europe and European Russia controls the "heartland." Who rules the heartland controls Eurasia. Who rules Eurasia controls the world. The path to global domination thus starts in Eastern Europe and European Russia. In the early twentieth century, naval warfare dominated military strategy. A land-based location, such as Eastern Europe and European Russia, would be relatively immune to naval attack. If you can consolidate your power and control in such a location you can then expand outward. After the end of the Second World War the Soviet Union was in control of all of Russia, Eastern Europe, and much of central Asia.

In the mid-1900s, an American researcher, **Nicholas Spykman**, introduced an alternate theory of geopolitics: the **rimland theory**. According to Spykman, who rules the fringe of Europe, Africa, and Asia controls the rimland. Who rules the rimland controls Eurasia. Who rules Eurasia controls the destiny of the world.

Subregions of Russia

We will examine two subregions of the Russia region: the Russian Core and Transcaucasia.

The Russian Core

The **Russian Core** spatially coincides with the Russian Plain, the European part of Russia. Most of the population, industry, and agriculture are located here. Within the Russian Core, the focus is the Central Industrial District. The **Central Industrial District** is based on the city of Moscow, the political, economic, and population center of Russia. Under the former Soviet Union, all decisions originated and diffused from the Moscow area.

The **North Caucasus** is that part of the Russian Core on the north side of the Caucasus Mountains. It is politically still part of the country of Russia. In the mind of ethnic Russians, the true border of Russia is culturally and politically the Caucasus Mountains. The North Caucasus includes a number of territories such as Chechnya and Dagestan. This is an area of Russia that includes mostly non-Russian ethnic groups who tend to be dominantly Muslim, such as the Dagestanis and the Chechens. The north Caucasus is an area of intense political and ethnic instability within Russia. Intense nationalism and the dominance of Islam provide the basis for much of the conflict in this area, especially in Chechnya.

Transcaucasia

Transcaucasia refers to the part of this region south of the Caucasus Mountains. Transcaucasia used to be part of the Soviet Union. It now includes three politically independent states: Georgia, Armenia, and Azerbaijan.

In **Georgia**, Russia recently invaded two areas with the intent of assisting ethnic Russians. These two areas are **Abkhazia**, in northwest Georgia, and **South Ossetia**, in north central Georgia.

Armenia has been in conflict with Azerbaijan over an area known as Nagorno-Karabakh. **Nagorno-Karabakh** is a territory within the political borders of Azerbaijan but comprised of mainly Armenians. Armenia has been trying to unite the Armenians in Nagorno-Karabakh with the rest of Christian Armenia. Muslim-dominated Azerbaijan resists the loss of any of its territory. Religion and ethnicity provide the basis for much instability between these two countries.

Azerbaijan, with newly discovered oil deposits off its Caspian Sea coast, is an emerging major oil producer within the subregion.

UNITED STATES AND CANADA

Where Is the United States and Canada?

As a region, the United States and Canada are the two northern countries of the continent of North America. Most of our focus will be on the United States.

Comparison and Contrast

The two countries are comparable in territorial size. Canada is slightly larger than the United States. Canada is the world's second largest country in land area and the United States ranks fourth in land area.

Both countries are mainly located in the middle latitudes. Since most of Canada is north of the United States it does include a larger area of colder, drier climate conditions.

Politically, the United States and Canada are federal states. A **federal state** is a country where power is distributed through different levels: national, state (province), county, and municipality. Both countries have a democracy-style government; Canada has a parliamentary democracy while the United States has a representative democracy.

The two countries are very economically developed. Both have high income levels, high production of manufacturing, and maintain high levels of trade with each other. They are both very highly urbanized.

Culturally, the United States and Canada are dominantly Indo-European countries that are primarily Christian in religion. English is the major language spoken in the United States though it is not an official language of the country. In Canada, there are two official languages: English and French. Most U.S. Americans are Protestant Christian. In Canada, Catholic Christians represent a higher share of the population than in the United States. Canada does have a stronger presence of European culture and lifestyle than the United States.

The biggest contrast between the two countries is population size. There are about 315 million people in the United States; Canada only has about 35 million people.

Landform Patterns

Topographically, river, interior, and coastal plains dominate the eastern two-thirds of the United States and the eastern four-fifths of Canada. In the United States, the Atlantic and Gulf Coastal Plain, the Interior Lowlands, and the Great Plains are examples. Mountains and high plateaus dominate the western parts of both countries. These include the parallel north to south trending Rocky, Sierra-Cascade, and Pacific Coast mountains. The Intermountain Plateaus lie between the Rocky Mountains and Sierra-Cascades.

Climate Patterns of the United States

Most of the United States has humid climates.

A **humid temperate (mild) subtropical climate** dominates in the southeastern U.S. There are long, hot summers and short, mild winters, with abundant precipitation through the year. New Orleans, Atlanta, and Charlotte are typical examples.

A **humid temperate (mild) Mediterranean climate** dominates in coastal central and southern California. Summers are warm and dry, with winters being mild and wet. San Diego, Los Angeles, and San Francisco are examples.

A **humid temperate (mild) marine climate** dominates in the Pacific Northwest. Summers are cool and wet, with winters being mild and wet. Conditions include clouds, fog, and misty rain for much of the year. Seattle and Portland, Oregon, are examples.

A **humid, cold (continental) climate** dominates in New England and the Great Lakes areas. Summers tend to be shorter and cooler with long, cold winters. Boston and Chicago are examples.

Dry climates dominate in the Great Plains, the Intermountain West, and the Southwest U.S. These areas have semi-desert to desert conditions, with less than 20 inches of annual precipitation. The dry conditions only support short grasses and desert shrubs. Phoenix, Las Vegas, and Salt Lake City are examples.

Rainfall tends to decrease as you travel westward from the Atlantic coast. Most of the East receives 40–80 inches; the central area receives 20–40 inches; the Great Plains receive 10–20 inches; and the Southwest deserts receive less than 10 inches. The main exception is the Northwest Pacific coast, which has a very mild and very wet climate.

Rain Shadow Effect

The side of mountains that directly faces oncoming air is the **windward** side. On the windward side air rises and cools, leading to wet conditions. The side of mountains opposite the flow of oncoming air is the **leeward** side. Air descends, warms, and dries out on the leeward side. The **rain shadow effect** refers to the warm and dry air on the leeward side of mountains.

Given the north to south orientation of U.S. and Canadian mountains, the west side tends to be the windward side; the east side tends to be the leeward side.

Humid East versus Dry West

The **100th meridian** (line of 100 degrees west longitude) is a significant boundary line for Canada and the United States, separating humid conditions on the east from dry conditions on the west.

The **Dry West is** located west of the 100th meridian; averages generally less than 20 inches of annual precipitation; includes mostly short grasses and desert shrubs; has a browner landscape; and has a sparser and more scattered population.

The **Humid East** is located east of the 100th meridian; averages more than 20 inches of annual precipitation; includes mostly tall grasses and forests; has a greener landscape; and has a more dense and concentrated population.

Native American Arrival and Impacts

Evidence indicates that the first Native Americans entered the Americas anywhere from 10,000 to 100,000 years ago. This time period coincides with the latter part of the Pleistocene Ice Ages. At this time a land connection, the **Bering Land Bridge**, connected Asia to Alaska. The Native Americans came from northeast Asia and entered by this land bridge into central Alaska, which was an unglaciated area at the time. Material (fossil) evidence and nonmaterial (language) evidence indicates the time of arrival of the Native Americans.

The Native Americans had a similar ecological philosophy to most Asians—a respect for the environment and a desire to maintain harmony and balance with the environment. Thus, few changes were made in the physical environment.

In the east, Native Americans were mainly farmers who needed to clear the forests for fields and settlements. They used a **slash-and-burn** method to cut and burn the forests, creating **old fields** (open areas within the forest). In the central areas, slash-and-burn farming changed former natural forest areas into tall, lush **prairie grasses**. The Native Americans used natural features of the landscape to create odd-shaped, irregular plots of land—the **metes and bounds survey system**. Indian trails followed the irregular field boundaries, creating a curving, irregular set of pathways. Only a few animals (turkey, guinea pig, llama, and alpaca) were bred for human purposes. Many crops originated with the Native Americans—maize (corn), tobacco, squash, beans, tomato, potato, and pineapple. The most visible impact of the Native Americans was in place names (**toponyms**)—Chesapeake, Potomac, Patapsco, Patuxent, Mississippi, Missouri, Utah, and Dakotas are examples.

European Arrival and Impacts

Written evidence indicates that Irish missionaries were the first Europeans to arrive in the Americas, as early as 700 AD. They were few in number and made no lasting impact.

Scandinavians arrived by 1000 AD in larger numbers and left behind the oldest European settlement in the Americas, L'Anse Aux Meadows, on the island of Newfoundland.

The year 1492 represents not the first European contact, but the beginning of major European settlement and influence.

Similar to Native Americans, a major landscape impact by the Europeans was in the form of toponyms—Maryland, Virginia, Baltimore, Annapolis, New Orleans, Los Angeles, San Francisco, New York, New Jersey, Bronx, and Yonkers are a few examples. The Europeans had a different ecological philosophy than the Native Americans. Most Europeans believed that the environment was there to be used to the maximum possible. Thus, in a few centuries, Europeans made extensive changes to the physical environment. The Europeans also brought many new infectious diseases that led to a rapid decline in many of the Native American populations.

Population Distribution and Change

The 350 million people in the United States and Canada region are very unevenly distributed. Most of the U.S. population occupies the humid East while most of Canada's population is in the warmer, southern areas (most within 100 miles of the U.S. border).

At the beginning of the twentieth century, most Americans lived in the Midwest and South census regions, with the smallest number occurring in the West census region. By the early twenty-first century, the South census region was the most populated with the West becoming second most populated. These changes are the result of internal population movements from east to west, north to south, and rural to urban.

There are three main periods of external migration to the United States: pre-1890, 1890–1914, and post-1950.

Prior to 1890, most migrants were European and overwhelmingly from northern and western Europe—British, German, and Scandinavian. Thus, most migrants were WASPs (White Anglo-Saxon Protestants), linguistically, religiously, and physically similar people.

The period of **1890–1914** is known as the **Great Deluge**. Until recently, this period was the largest in migrant size in U.S. history. About 25 million Europeans migrated to the United States, most of them from southern and eastern Europe—Italians, Poles, Czechs, and Slovaks are principal examples. This newer group was physically, linguistically, and religiously different from the earlier migrants. Intense ethnic and religious discrimination occurred against these "foreigners."

From 1914–1950, World War I, the Great Depression, World War II, and European recovery halted most European migration to the United States.

Since 1950, the pace of migration quickened but the sources became quite different. Currently, the size of migration matches the Great Deluge. Europe ceased to be a major source area and was replaced by Latin America (Mexico, Central America, and the Caribbean) and Asia (Philippines, Vietnam, China, and India are main sources).

U.S. Religious Regions

There are five main religious regions that dominate the United States: Roman Catholic, Baptist, Lutheran, Methodist, and Mormon.

The **Roman Catholic religious region** includes two widely separated areas: the Northeast and the Southwest. Originally, the Northeast was a Protestant Christian area. Beginning in the 1830s and 1840s, major Irish migration began to cities like Boston, New York, and Philadelphia. Then, larger numbers of Italians and eastern Europeans arrived during the Great Deluge to similar big cities of the Northeast. Over the last 150 years the Northeast became a Catholic-dominated area. Spanish missionaries and explorers introduced Catholicism to the Southwest in the 1500s and 1600s. More recently, huge numbers of Mexicans have intensified the Catholic dominance in the Southwest.

The **Baptist religious region** includes most of the Southeast United States. British migrants introduced Baptist Protestantism. Baptist missionaries on horseback rapidly spread this denomination throughout the South.

The **Lutheran religious region** includes most of the northern Great Plains and upper Midwest (Dakotas, Minnesota, Iowa, and Wisconsin). Migrants from northern Germany and Scandinavia introduced Lutheranism into this area.

The **Methodist religious region** extends from Maryland westward into the central Midwest. British migrants introduced Methodist Protestantism into central Maryland and it then spread directly westward.

The **Mormon religious region** includes the Intermountain West (the area between the Rockies and the Sierra-Cascades). The Church of Jesus Christ of Latter Day Saints (the LDS or Mormon religion) originated in upstate New York in 1830. The Mormons migrated to the Utah area in the late 1840s. They then rapidly spread their influence throughout most of the Intermountain West. Though they are Christian, they do not belong to either the Catholic, Protestant, or Orthodox branches.

Racial and Ethnic Groups in the United States

At the end of the twentieth century, the majority racial/ethnic group was non-Hispanic White. The largest minority group was Black or African American. The Hispanic population was the second largest minority group, followed by the Asian and Native American groups.

As of the 2000 census, the Hispanic population became the largest minority group, slightly larger than the Black or African American population. The Asian population was the third largest minority group and the smallest minority group was the Native American.

By 2050, the non-Hispanic White population will be a slight majority. The Hispanic population will be clearly the largest minority group; the Black or African American population will be the second largest; the Asian population the third largest; the Native American population will continue to be the smallest group.

Minority Group Patterns of the United States

The **Hispanic** population is now the largest minority group. Hispanics are a very diverse ethnic group. The largest Hispanic group is Mexican, which dominates in the U.S. Southwest. The Cuban group dominates in south Florida. The Puerto Rican group is dominant in the New York City area. New Mexico has the highest proportion of its population that is Hispanic. California has the largest number of Hispanics.

The **Black or African American** population dominates in the Southeast U.S. Most of this group is descended from slaves brought to the Southeast to work on plantations. Mississippi has the highest proportion of its population that is Black. New York State has the largest number of Blacks.

The **Asian** population is mainly evident in the Pacific coast area. The Asians are another very diverse group; Chinese, Filipinos, Vietnamese, and Asian Indians are examples. Hawaii has the highest proportion of its population that is Asian. California has the largest number of Asians.

The **Native American** population is mainly found in the Western U.S. The major area of dominance is the Four Corners area (where Utah, Arizona, Colorado, and New Mexico meet). Navajo, Hopi, Zuni, and Ute groups are concentrated in this area. Alaska has the highest proportion of its population that is Native American. California has the largest number of Native Americans.

Economic Changes and Patterns

Up until the late 1800s the United States was mainly an agricultural economy. Beginning in the late 1800s, the U.S. shifted toward an industrial economy. By 1950, the U.S. became a **post-industrial economy**, mainly comprised of marketing and service activities.

The **North American Manufacturing Belt** is a large rectangular-shaped area that extends from Boston on the Northeast to Baltimore on the Southeast to St. Louis on the Southwest to Milwaukee on the Northwest. The greatest concentration of U.S. industrial activity and industrial workers occurs in this region. Over time, regional specializations developed—automobiles in the Detroit area, steel in the Pittsburgh-Chicago area, beer in the Milwaukee and St. Louis areas; finance, garment, and publishing in New York City.

Several agricultural regions still exist in the United States. Cotton and tobacco dominate in the southeast. A **Dairy Belt** dominates in the Northeast and Great Lakes areas. The **Corn Belt** is a collection of corn, soybeans, cattle, and hogs that dominate in the central area, especially Illinois and Iowa. The **Wheat Belts** dominate in the Great Plains. **Truck farming** (fresh vegetables and fruits trucked into nearby large urban areas) dominates along the Middle Atlantic coast from Virginia to Long Island.

U.S. Subregions

We will examine four subregions within the United States. These include the North American Core, the Continental Interior, the Southeast, and the Southwest.

North American Core

The **North American Core** is spatially identical to the North American Manufacturing Belt. The greatest concentration of people, industry, big cities, and economic activity occurs here. It is not the most self-sufficient area since it must import most of its mineral and food resources. In the 1950s, a visiting French geographer, Jean Gottmann, coined the word "megalopolis" to refer to a continuous series of overlapping urban areas. Conceptually, **megalopolis** refers to any place where two or more metropolitan areas overlap. Examples of megalopolis include Chipitts (Chicago to Pittsburgh), SanSan (San Diego to Santa Barbara), and Houville (Houston to Jacksonville). As a region, **megalopolis** refers to the densely populated North Atlantic coast, from the northern suburbs of Boston in southern Maine to the southern suburbs of Washington in northern Virginia (also known as **Boswash**).

The Continental Interior

The **Continental Interior** includes the central interior section of the United States. This is the most self-sufficient area of the U.S. as it includes most of the food production, many minerals, lots of cities, and a relatively large population. Over time, several important images have been used to depict this area. The Continental Interior is often referred to as the "**heartland.**" This heartland image refers to the centralized location of this subregion and its vital basis for the existence of the overall region. Another image of this subregion is that of the "**breadbasket.**" The Wheat Belts and the Corn Belt occur in the Continental Interior, providing most of the food production of the country. The final image is "**Main Street.**" The Continental Interior is comprised of mainly small towns, with their Main Street business area. Walt Disney used this image as a centerpiece of his Walt Disney World theme park.

The Southeast

The **Southeast** includes the area extending from Texas to Maryland. The Southeast is a very dynamic area, with many changes. We need to distinguish between an Old South and a New South.

The **Old South** refers to the Southeast prior to 1950. The Old South was dominantly a rural, agricultural, less developed area. Plantation agriculture, slave labor, low incomes, high levels of poverty, and outmigration of people characterize the Old South. Atlanta is the historical capital of the Old South.

The **New South** refers to the Southeast since 1950. The New South is dominantly an urban, industrial, rapidly growing area. Diversified agriculture, big cities, rising incomes, in-migration of people and businesses, and a fast-growing economy characterize the New South. Interestingly, Atlanta is the capital of the New South.

The Southwest

The **Southwest** includes the area from the Rio Grande Valley to southern California. Similar to the Southeast, the Southwest is also a very dynamic area. There has been rapid population and economic growth, caused by internal and external influences. The Southwest is known as a **tricultural area**, influenced by three major historical groups: Native Americans, Hispanic Americans, and Anglo Americans.

Native Americans are the original occupants of this area, with their greatest concentrations in the Four Corners Area. The largest Native American group, the Navajos, dominate much of this area. There are also Hopis, Zunis, and Utes. Most of the Native Americans live on reservations, with high levels of poverty, alcoholism, and crime.

Hispanic Americans include the early Spanish and the Mexicans. San Antonio represents the historical cultural center of this group. It was the first U.S. city to elect a Hispanic mayor.

The **Anglo Americans** are the most recent group, bringing northern and western European influences to this area.

These three cultural groups have not interacted very well, leading to many conflicts.

MIDDLE AMERICA

Where Is Middle America?

Middle America refers to the North American continent south of the United States. It is a section of the broader area known as Latin America. **Latin America** refers to all of the Americas south of the United States, including Middle America and South America.

As a region, Middle America includes Mexico, Central America, and the Caribbean island states.

Major Qualities of Middle America

Middle America is physically and politically fragmented. Physically, it includes countries on the North America mainland as well as in the Caribbean Sea. Politically, there are a large number of individual countries.

The mainland is a land bridge between North and South America, and acts as a barrier between the Atlantic and Pacific Oceans.

Middle America is culturally, racially, and ethnically very diverse. There are several major groups that comprise the complex ethnic mosaic of this area.

There are approximately 200 million people who live in Middle America, with the majority of the population in one state, Mexico.

Middle America is a less-developed economic region, with widespread poverty, low incomes, and relatively rapid population growth. Recently, some economic reforms and increasing industrialization have occurred, especially in Mexico.

Physical Geography

Middle America is comprised of a mainland section that is a land bridge between North America and South America. There is also a rimland section that includes two groups of Caribbean islands: the Greater Antilles and the Lesser Antilles. The **Greater Antilles** includes the four largest and most populated Caribbean islands: Cuba, Jamaica, Hispaniola, and Puerto Rico. The **Lesser Antilles** includes the smaller and less populated Caribbean islands, such as Martinique, Barbados, U.S. Virgin Islands, Netherland Antilles, and Grenada.

Middle America lies in the tropical and subtropical latitudes, with humid equatorial (tropical) climates predominating. The location is a major focus of hurricane activity, with hurricanes either originating in the Caribbean Sea or moving into the area from the Atlantic Ocean. The west coast of Mexico is also affected by hurricanes coming from the Pacific Ocean.

In addition to the weather hazard of hurricanes, there are also many geologic hazards. Middle America's location places it at the intersection of a number of major earth plates, including the Pacific Ring of Fire. Many earthquakes and volcanoes affect the Middle America region.

Major Altitude Zones of Latin America

Four major altitude zones exist in the mountainous areas of Middle America and South America: tierra caliente, tierra templada, tierra fria, and tierra helada.

The **tierra caliente** is a hot and wet zone that occurs from sea level to about 3,000 feet. It is very warm and very wet, not climatically preferable for Europeans. This is a zone of tropical plantation agriculture, including sugar, bananas, rice, and other tropical crops.

The **tierra templada** is a temperate zone reaching to about 6,000 feet. Latin America's largest population concentrations occur in this intermediate area. Grain crops such as maize (corn), wheat, and many vegetables are grown in this zone as well as commercial coffee production.

The **tierra fria** is a cold area extending to about 12,000 feet. Crops such as potatoes and barley predominate in this zone.

The **tierra helada** is the fourth altitude zone, extending to about 15,000 feet. This altitude is above the tree line and it can support only sheep grazing and other livestock herding on a seasonal basis.

Tropical Deforestation

A major environmental issue in Latin America is tropical deforestation. **Tropical deforestation** is a process of forest clearance, mainly resulting from human activity. The greatest degree of global tropical deforestation occurs in Middle America and South America.

There are multiple causes of tropical deforestation. Agriculture is one important factor. In order to plant and grow crops, tropical forests need to be cleared. The most common way of removing the trees is through a practice known as slash-and-burn agriculture. **Slash-and-burn agriculture** includes people slashing the trees in one season and then burning the debris in another season. Cattle grazing is a rapidly growing human activity in Middle America. Many tropical farmers are aware of a large market in the developed countries for beef. They can make more money grazing cattle than growing crops. However, trees still need to be removed for grazing to take place. In addition to farming and grazing, settlement of people in towns and villages requires forests to be cleared. Road construction also is an important cause of deforestation. The tropical forests include commercially valuable species such as rubber, mahogany, teak, and ebony. However, these tree species are intermixed with other trees. To obtain one rubber tree may require extensive clearing of many other trees. Logging activity has become a significant cause of tropical deforestation.

There are also multiple consequences of tropical deforestation. The biosphere is adversely affected. In addition to the loss of the trees, habitat for many animals and plants is reduced or lost, leading to the possible disappearance of the species themselves. With tree removal, much precipitation is no longer absorbed by tree roots and leaves. This leads to increased water runoff and erosion of soil. The soil then winds up in streams, causing sedimentation. The fast runoff of the water leads to flooding of the stream. The chemistry of the atmosphere also changes. With fewer trees, there is less oxygen released and more carbon dioxide remains in the atmosphere. The increase in atmospheric carbon dioxide intensifies greenhouse warming. Warming is also intensified by the burning of the trees; burning releases even more carbon dioxide. Increasing deforestation also allows more solar energy to reach the surface, leading to even more warming.

Early Indian Civilizations

There are two major pre-European Native American groups in Middle America: the Maya and the Aztecs.

These two civilizations included dense concentrations of people in cities with monumental architecture; early agriculture; mathematical concept of zero; architectural notion of the arch; and presence of writing.

The **culture hearth** (place of origin) **of the Maya** is the lowland tropics of southern Mexico and the Yucatan Peninsula. Maya culture reached its pinnacle from the third to the tenth centuries AD.

The **culture hearth of the Aztecs** is the highland zone of central Mexico, around what is today Mexico City. Aztec culture lasted from the fourteenth century to the sixteenth century.

Early European Colonialism

The Spanish were the dominant Europeans in the mainland area of Middle America. In the Caribbean, Spanish colonialism occurred in Cuba, Dominican Republic, and Puerto Rico.

The British had extensive colonial holdings in the Caribbean, including Jamaica, Bahamas, British Virgin Islands, Grenada, Barbados, and Trinidad and Tobago. On the mainland, the British controlled British Honduras (now Belize).

The French had colonial control in Caribbean states such as Haiti, Guadeloupe, and Martinique.

The Dutch had colonial possessions in the Netherlands Antilles (Aruba, Bonaire, and Curacao).

U.S. colonies included Puerto Rico and the U.S. Virgin Islands.

Cultural and Racial Diversity

There are five main cultural/racial groups that comprise the population of Middle America: Native American Indian, European, Mestizo, African, and Mulatto.

Native American Indians are the original occupants of Middle America. Today, they are dominant in southern Mexico, the Yucatan Peninsula, and Guatemala.

Spanish Europeans historically dominated in Mexico and most of Central America. In the Caribbean islands, there were a variety of European groups, including Spanish, French, British, and Dutch. Today, the country of Costa Rica is dominantly European.

Mestizo is a blended group, with mixtures of Indian and European ancestry. The mestizo group dominates in Mexico and most of Central America.

African slaves were imported to the Caribbean islands and coastal areas of Central America to work on plantations. Africans today dominate in the Caribbean islands.

Mulatto is a blended group, with mixtures of European and African ancestry. The mulatto group dominates in the Caribbean islands and in the coastal parts of Central America.

Haciendas

Haciendas are land use systems comprised of large landholding units owned by a few wealthy individuals. Most of the people are peasants who work on the hacienda. The Spanish introduced this land system into the mainland area of Middle America. Haciendas are a mixed type of farming, with diversified crops grown for domestic use and extensive cattle grazing. It is not a very efficient system but primarily based on owning large areas of land with lots of cattle for social prestige purposes.

Plantations

Plantations are land use systems comprised of large landholdings owned by a few wealthy people. Most of the work is done by slaves imported to work on the plantations. Northwest Europeans introduced this land system into the Caribbean islands and Caribbean coastal parts of Central America. Plantations are a type of monoculture, where one cash crop is grown for export purposes. Commercial profit is the main purpose. Coffee, sugar, rice, and bananas are common plantation crops.

Mainland and Rimland

One common way of dividing Middle America is into the Mainland and the Rimland areas.

The **Mainland area** is comprised of Mexico and most of Central America. It is mainly a Euro-Indian dominated area, with primarily Spanish European and Native American Indian group influences. The hacienda land use system dominates the economy.

The **Rimland area** is comprised of the Caribbean islands and a narrow Caribbean coastal strip of Central America. It is mainly a Euro-African dominated area, with a variety of diverse European and African influences. The plantation land use system dominates the economy.

Middle America Subregions

Middle America is comprised of three subregions: Mexico, Central America, and the Caribbean.

Mexico

Mexico is the dominant country in Middle America. It has approximately 115 million people, over half of the region's total population. Mexico is the largest country in land area and is the leading economic country within Middle America. Mexico's large land area provides many mineral resources, including oil, coal, natural gas, silver, iron, copper, and gold.

The **Core Area of Mexico** extends from Veracruz on the Gulf of Mexico coast to Guadalajara on the west, with Mexico City in the middle. Mexico City is the largest city, political capital, and economic center of Mexico.

Mexico's southern region and the Yucatan peninsula include the country's largest concentration of Native American Indians. The southern region tends to be economically poor with much political instability.

Maquiladoras are factories owned by U.S., Japanese, and European companies that are concentrated along the northern border with the U.S. They represent some of the fastest growing economic areas within Mexico. The incentive for the foreign companies is the attraction of cheap wage labor. Many U.S. companies have moved much of their basic assembly operations into northern Mexico.

NAFTA (**North American Free Trade Agreement**) went into effect in 1994. It established a free trade agreement among Canada, the United States, and Mexico that was intended to allow for the easy movement of goods among these three countries. It has greatly benefitted Mexico, leading to even more U.S. factories relocating to Mexico for its cheaper labor costs. This has resulted in a greater loss of U.S. jobs to Mexico. U.S. labor unions have greatly opposed the agreement. Many environmental organizations have also opposed it due to the lack of pollution laws and regulations in Mexico.

Central America

Central America includes the mainland countries between Mexico to the north and Colombia to the south. These countries include Guatemala, Belize, Honduras, El Salvador, Nicaragua, Costa Rica, and Panama.

Guatemala is the only predominantly Native American Indian country in Central America. It is economically very poor and politically unstable.

Belize and El Salvador are locationally different from the other Central America countries due to Belize only having frontage on the Caribbean Sea and El Salvador only having frontage along the Pacific Ocean. Belize is also distinctive in being a former British colony, whereas Spain dominated most of the rest of Central America.

Costa Rica is a political democracy, very rare in Central America. It has almost a purely European population, with higher levels of economic development relative to the rest of the subregion. It has no standing army, and has the largest proportion of its territory devoted to national parks. Costa Rica is globally known for its ecotourism industry. **Ecotourism** is the process by which a country uses its natural ecology as a tourist draw. Costa Rica has a very high percentage of its budget devoted to creating and maintaining ecological areas that are used to attract tourists.

The Caribbean

The **Caribbean** includes two sets of islands: the Greater Antilles and the Lesser Antilles.

The **Greater Antilles** are comprised of the four largest and most populated islands—Cuba, Jamaica, Hispaniola, and Puerto Rico.

Cuba is the largest and most populated Caribbean country. It was a longtime Spanish colony before its independence. Since the late 1950s, it has been controlled by a Communist government led by the Castro brothers, Fidel and Raul.

Jamaica, to the south of Cuba, is a former British colony. It is known for its bauxite (aluminum ore) resources as well as reggae music and Bob Marley.

Hispaniola, to the east of Cuba, is an island now politically divided between Haiti on the west and Dominican Republic on the east. **Haiti** is a former French colony, with historical dictatorships, extensive poverty, and low incomes. Haiti is generally considered to be the poorest economy in the Western Hemisphere. Much of the country has been heavily deforested, leading to lots of mudslides. Haiti is also prone to earthquakes and hurricanes. **Dominican Republic** is a former Spanish colony, now known as a major source of professional baseball players.

Puerto Rico is located east of Hispaniola. It is the easternmost and smallest of the Greater Antilles. Puerto Rico was a Spanish colony until the end of the Spanish-American War in 1898, when it became an American colony. In 1948, Puerto Ricans elected their own governor. In a 1952 referendum, voters approved the creation of a Commonwealth associated with the U.S. Puerto Ricans are U.S. citizens but pay no federal taxes on local incomes. There are many tax incentives for mainland companies and annual subsidies from the U.S. government.

SOUTH AMERICA

Where Is South America?

South America is identical to the continent of South America. It is the southern part of the broader area known as Latin America.

As a region, South America includes the North, the West, the South, and Brazil.

Major Qualities of South America

The Andes Mountains and the Amazon Basin physically dominate South America.

There are approximately 400 million people in South America, with half of the population in one state, Brazil.

South America is culturally, racially, and ethnically very diverse. The same five groups that comprise the population of Middle America also create a complex ethnic mosaic of this region.

South America is a less-developed economic region with huge disparities between rich and poor, leading to much political and cultural instability. Urbanization is very high in South America despite its being a less-developed region.

Physical Geography

The **Andes Mountains** are the longest continuous mountain range on the earth's surface. They extend north to south along the western edge of the continent. Mount Aconcagua lies in the southern part of the Andes, along the border of Chile and Argentina. **Mount Aconcagua** is the highest peak in the Americas, taller than Mount Denali (formerly known as Mount McKinley) in Alaska. Within the central Andes, especially in Bolivia, is the Altiplano. **The Altiplano** refers to a series of mountain valleys in the Andes that allow a dense concentration of people and activity.

There are two main grassland environments in South America: the llanos and the pampas. The **llanos** are a lowland grassy area astride the border of Colombia and Venezuela. The **pampas** are a lowland grassy area in eastern Argentina. The pampas are home to the gauchos (the Argentinian equivalent to the U.S. cowboy).

The **Amazon Basin** is an extensive tropical rainforest whose watershed occupies half the land area of South America. The world's largest area of tropical deforestation occurs in the Amazon Basin. The Amazon River is the world's largest river in terms of volume of water and size of watershed area.

There are two desert environments in South America: the Atacama Desert and Patagonia. The **Atacama Desert** is a tropical desert in northern Chile. It is the driest desert in the world. A combination of a semi-permanent high pressure system; a cold ocean current offshore; a position on the leeward side of South America; and a leeward location on the rain shadow side of the Andes led to very dry conditions here. **Patagonia** is a middle latitude desert in southern Argentina. At its location it is on the rain shadow side of the Andes.

The Inca Empire

The **Inca** was the major pre-European Indian group in South America. Their culture hearth was concentrated in the Andean Altiplano in the area of Bolivia. Similar to the Aztecs and the Mayans, the Incas were an advanced civilization with monumental architecture, the mathematical concept of zero, and the architectural idea of the arch. They were a highly centralized state with rather impressive transportation networks. Their peak occurred around 1300 AD.

The Tordesillas Line

In 1494, the then-current Pope used a line of longitude to separate the Americas between the Spanish and the Portuguese. This line is the **Tordesillas Line** and was drawn as a result of the Treaty of Tordesillas. Spain was granted areas west of the line while Portugal was awarded areas east of the line.

Cultural and Racial Matrix

There are five main cultural/racial groups that comprise the population of South America: Native American Indian, European, Mestizo, African, and Mulatto.

Today, Native American Indians are concentrated in the Andean Mountains. Europeans dominate the southern states of Chile, Argentina, and Uruguay. They also are dominant in southern Brazil. Mestizo is the largest group, dominating in countries such as Brazil, Colombia, Venezuela, Peru, Ecuador, and Paraguay. African and Mulatto groups are mainly found in the coastal areas of northern South America and northeast Brazil.

Geography of Cocaine

Most coca plant production occurs in Bolivia, Peru, and Colombia. Cocaine is then distributed to Mexico with the ultimate market being the U.S. About 50 percent of the cocaine reaches Mexico from South America along the Pacific coast. About 25 percent comes to Mexico overland from South America through Guatemala. Another 25 percent comes to Mexico from South America along the Gulf of Mexico coast. About 90 percent of the cocaine enters the U.S. from Mexico. Inside Mexico the cocaine is controlled by a series of drug cartels that dominate many areas of Mexico.

The Latin American City

Urbanization rates are very high in both Middle America and South America. These high urban rates are largely a consequence of the Spanish and Portuguese influence.

The spatial structure of the Latin American city is quite different than the North American city. In Latin America, the center of the city has always been perceived to be a place of status and prestige. The wealthy groups tended to concentrate near the center to take advantage of its location relative to cathedrals, palaces, and museums. The outskirts of the Latin American city are where most of the landless poor live in squatter settlements. In contrast, the North American poor minority groups are concentrated adjacent to the business district with the middle and upper income groups dominant on the outskirts in suburban areas.

South America Subregions

South America is comprised of four subregions: The North, The West, The South, and Brazil.

The North

The North includes the countries of Colombia, Venezuela, and the Guianas. These countries are located on the Caribbean and Atlantic coast. Economically, these very warm and wet tropical climates were very conducive to early European plantation development. The general lack of Native American Indians led the Europeans to import forced laborers from Africa and Asia.

Colombia is the most populated country in The North subregion. It is mainly known for its production and export of coffee and cocaine. Colombian coffee is well known for its high quality. Cocaine production has been the basis for much of the violence and instability within Colombia over the last few decades.

Venezuela is economically known for its oil production and exports. In the late 1990s Venezuela elected an anti-American, anti-free trade candidate, Hugo Chavez. Under Chavez and his successor, Venezuela became an autocratic state with mainly leftist tendencies.

The Guianas are three small and less populated countries along the northeast coast of South America: Guyana, Suriname, and French Guiana. Most of the people in the Guianas are African and mulatto. These are the only areas of South America that are neither Spanish nor Portuguese in language and settlement.

Guyana is a former British colony that once was known as British Guiana. English is still an official language. Guyana is most identified with the Jonestown Massacre in the late 1970s. Hundreds of members of an American religious cult died from mass suicide/murder under the leadership of a charismatic leader named Jim Jones. Gold and oil are major economic resources.

Suriname is a former Dutch colony that used to be known as Dutch Guiana. Dutch is still the official language. Suriname includes relatively large Asian groups in its population. Bauxite (aluminum ore) and oil are important economic minerals. Both Guyana and Suriname have significant issues in terms of tropical deforestation.

French Guiana is South America's only dependency. It is officially a department of France and French is the official language. Gold and fishing are traditional resources. The European Space Agency has a launch complex on the coast which accounts for the majority of the economic activity now.

The West

The West is economically the poorest, least developed subregion of South America. The Andes Mountains topographically dominate the area. Historically the Native American Indians have been the dominant occupants of the land. The West includes the countries of Ecuador, Peru, and Bolivia.

Ecuador is the smallest of the three Andean states. Its name comes from its location astride the equator. Guayaquil is the largest city, the commercial center, and the main port of Ecuador. The capital is Quito, in the Andean mountains. Oil is the main export, with bananas being a traditional plantation crop.

Peru is the largest country in area and population within The West subregion. Peru has an extensive set of mineral resources, including copper, lead, and gold. Recent oil and natural gas discoveries in the East may help to stimulate development of this part of the country. Peru is most known for Machu Picchu, one of the best pre-European sites in South America. Near Cuzco, it was part of the hearth area of the Incan Empire.

Bolivia is one of two landlocked states in South America (Paraguay is the other state). It has been a major producer of coca and cocaine that is eventually transported to Mexico and the U.S. There is a significant European and Native American Indian divide within Bolivia. The western, mountainous part is dominated by Native American Indians, who are the majority of the population. The coca producing areas are in the West as well as Bolivia's two capitals. La Paz is the administrative capital while Sucre is the judicial capital. Bolivia has the only elected Native American Indian president, Evo Morales. The eastern lowlands are dominated by the Mestizo minority, actively seeking more autonomy for their part of the country. Most of the mineral resources are in the east.

The South

The South is the only middle latitude part of South America. It is comprised of the largest proportions of Europeans in the continent. The South includes three countries: Argentina, Chile, and Uruguay.

Argentina has the most population and the largest land area of The South subregion. It is one of the most urbanized countries in the world; over 90 percent of the population lives in urban areas. Buenos Aires is a classic example of a primate city: it is the largest city, the political capital, and the economic center of the country. Argentina is a major exporter of beef and wheat. Argentina has always considered itself to be the most sophisticated state in South America. Nevertheless, Argentina did go through a turbulent period known as "The Dirty War" (mid-1970s to mid-1980s). A repressive military junta led to the disappearance of tens of thousands of young people and launched a failed invasion of the British-held Falkland Islands.

Chile is an elongated state that extends over 2,500 miles north to south. This extended latitude location provides dramatic differences in climate from the Atacama Desert in the North to a Mediterranean-type climate in the middle to a Marine climate in the South. Overwhelmingly, most of the population is concentrated in the milder central area, including the capital and largest city of Santiago. Fruit, wine, and vegetables are main exports from the central area. Copper is a main export from the northern Atacama Desert area. The world's largest open pit copper mine (the second deepest in the world) is within this northern area. The southern area is based on logging and timber as exports. Chile's agricultural economy benefits from its southern hemisphere location (Chile's summer season is the winter season for U.S. and Europe). Chile also benefits from its Pacific coast location with access to the growing Pacific Rim economies. One major problem of the Pacific Ocean location is that Chile lies on the Ring of Fire, with extensive major earthquakes often affecting the Santiago area.

Uruguay is a small, low population state across the water from Buenos Aires in Argentina. Montevideo is another classic primate city. Uruguay acts as a buffer between Brazil to the north and Argentina to the south. Meat products and textiles are important exports.

Brazil

Brazil is the dominant country in South America. It contains half the area and half the population. There are approximately 200 million people in Brazil (#5 in the world). The world's largest tropical rainforest, experiencing tremendous rates of deforestation, lies within the vast Amazon Basin. Brazil has one of the most diverse populations in South America. It is a mixture of European, African, and Native American Indian peoples that blend into a composite Brazilian culture. Brazil is the only country in South America that is Portuguese-speaking. A benefit of its large area size (#5 in the world) is a vast array of minerals, including bauxite, iron, oil, and gas. Historically, Brazil had the greatest disparity in income between rich and poor in the world, a contrast vividly seen in photographs today.

We will examine five areas of Brazil: The Northeast, The Southeast, Sao Paulo State, The South, and The Interior.

The **Northeast area** includes the cities of Salvador, Recife and Fortaleza. It is the country's culture hearth, the place of origin of Brazilian culture. Historically the plantation economy dominated the Northeast area; it contains the greatest concentrations of African and Mulatto groups. Today, it is economically poor and depressed, with extensive outmigration of people and activities toward the more populated and developed southern areas.

The **Southeast area** includes Rio de Janeiro, the former political capital and still cultural center of Brazil. Rio will be the location of the 2016 Olympics.

The **Sao Paulo State area** includes the city of Sao Paulo, the largest city in Brazil and South America. Sao Paulo is also one of the largest megacities in the world. Sao Paulo is the industrial center of Brazil. Its economy was initially stimulated by the success of coffee plantations (Brazil is #1 in world coffee production). Today orange juice concentrate (Brazil is #1 in world production) and soybeans (Brazil is #2 in world production) have become major economic leaders. Sao Paulo's population is a microcosm of Brazil: waves of Europeans (Portuguese, Italians, Germans, and British), Africans and mulattos from the Northeast area, and Japanese have created a rich urban tapestry. Today, Sao Paulo has the largest concentration of Japanese outside of Japan.

The **South area** includes that part of Brazil south of Sao Paulo. It is the most European, most developed, and wealthiest part of Brazil.

The **Interior area** includes the capital city of Brasilia. In 1960, Brazil built a modern, new city in the empty interior, built to reduce congestion and overcrowding along the southeast coast. Brasilia is a classic example of a forward capital. The Amazon Basin comprises the majority of land in this area.

SUB-SAHARAN AFRICA

Where Is Sub-Saharan Africa?

The continent of Africa consists of two widely different regions: North Africa and sub-Saharan Africa. North Africa includes the Mediterranean coast and the Sahara, the world's largest desert. **Sub-Saharan Africa** lies south of the Sahara.

Sub-Saharan Africa versus North Africa

There are significant climate, population, language, religion, and economic contrasts between North Africa and sub-Saharan Africa.

Dry climate conditions dominate North Africa, with most areas receiving less than 12 inches of annual precipitation. In contrast, humid climates dominate in sub-Saharan Africa, with most areas receiving 20–100 inches of precipitation annually.

Most of North Africa is very sparsely populated due to the largely dry climates there. Sub-Saharan Africa is much more populated, with dense concentrations in the west and east.

The Afro-Asiatic language family predominates in North Africa while the Niger-Congo language family dominates south of the Sahara.

Islam is the dominant religion in North Africa whereas Christianity and tribal religions dominate in sub-Saharan Africa.

Middle income countries predominate in North Africa while low income countries are dominant in sub-Saharan Africa.

Major Qualities of Sub-Saharan Africa

Africa, especially sub-Saharan Africa, is described as a plateau continent, with narrow coastal areas backed by steep escarpments with most of the land being a series of high plateaus.

Every country in Africa is a multinational, multiethnic state, largely resulting from inefficient political boundaries created by European colonial empires. The culturally diverse political landscape leads to lots of ethnic and religious conflicts causing huge numbers of dislocated peoples and refugees.

Sub-Saharan Africa is the world's least economically developed region, with low incomes and low economic output. Most people are engaged in subsistence farming, a hallmark of low development.

Over 800 million people inhabit the region and it is also the world's fastest growing area.

Another important characteristic of this region is that it has the world's highest concentration of endemic diseases, diseases that are chronic or native to an area.

Physical Geography

Sub-Saharan Africa is described as a **plateau continent**. Most of the land area is comprised of a series of high plateaus, bordered by steep escarpments that lead down to narrow coastal plains. Most rivers originate on the plateaus and then tumble over the escarpments in steep waterfalls with lots of rapids at the bottom. It is this distinctive topography that made sub-Saharan Africa a "dark" continent. To reach sub-Saharan Africa you either had to cross the world's largest desert or somehow ascend the steep escarpments. Europeans did not reach this area of the world until the late 1800s.

One of the most notable physiographic features seen from space satellites is the Great Rift Valley. The **Great Rift Valley** is a series of downfaulted basins that originate from plate tectonics and a landform process known as faulting. Faulting is a geologic process that involves the vertical shifting of crustal layers. In East Africa, faulting and plate movement created a set of valleys that filled with water to form a series of lakes and seas, such as Lake Victoria, Lake Malawi, the Red Sea, and the Dead Sea.

Africa's largest tropical lowland is the **Congo Basin**. The Congo Basin is Africa's equivalent to South America's Amazon Basin. It lies astride the equator and is an extensive tropical rain forest with very warm and very wet conditions.

The **Namib Desert** is a tropical desert within sub-Saharan Africa. It lies on the southwest coast of Africa, including most of the area of the countries of Namibia and Botswana.

The **Sahel** lies along the southern margins of the Sahara. It is a major transitional area with a mixing of varied climates, religions, languages, and economies.

The Bantu

The **Bantu** is the largest and most widespread language group within the Niger-Congo language family. The Bantu originated in western Africa and eventually dispersed throughout central and southern Africa, becoming the most extensive group in the region.

The Berlin Conference

Due to the plateau continent nature of sub-Saharan Africa, Europeans did not arrive in this region until the late 1800s. In 1884, most of the largest European countries met in Berlin and in the **Berlin Conference** they divided Africa into a series of European colonies. The Europeans were interested in acquiring important minerals and plantation crops. They used their economic motives to divide Africa without much consideration of native cultures, languages, and religions. These actions resulted in a distinctive set of political boundaries that separated the European colonies in Africa. **Superimposed boundaries** are boundaries drawn by outside interests with little regard to local cultures. These superimposed boundaries create largely multiethnic and multinational states that result in much political and cultural instability. Every country in Africa is a multinational state.

Sub-Saharan Africa Independence

Most countries within sub-Saharan Africa obtained their independence in the 1950s and 1960s. Ghana, in West Africa, was the first African country to get its independence from Europe after World War II.

Subsistence Agriculture

Subsistence agriculture is the dominant economic activity in sub-Saharan Africa. Most people are poor farmers who grow crops or graze animals for their own needs with little market orientation.

Medical Geography

The very warm and wet conditions that dominate most of sub-Saharan Africa are excellent environments for organisms like flies and mosquitos that carry and transmit many diseases. It is very appropriate then to investigate some of the main aspects of a subfield of geography known as medical geography. **Medical geography** studies the origin and diffusion of diseases.

Medical researchers commonly distinguish among three types of diseases: endemic, epidemic, and pandemic.

An **endemic disease** is chronic to a local area. It is omnipresent, physically weakening people exposed to it. Few people die from an endemic disease but it saps energy, lowers resistance, and helps to shorten lives. An endemic disease diffuses slowly leading to local concentrations. Hepatitis and hookworm are examples of endemic diseases in sub-Saharan Africa.

An **epidemic disease** suddenly outbreaks and spreads rapidly at a regional scale. Many deaths result. Sleeping sickness is an epidemic disease in sub-Saharan Africa.

A **pandemic disease** spreads rapidly and globally with many deaths. Malaria and HIV/AIDS are examples of pandemic diseases.

Research highly suggests that **HIV/AIDS** is a disease that likely originated in the eastern Congo River basin as an endemic disease. It probably was chronic to this local area over a long period of time. The HIV virus apparently began in chimpanzees and eventually spread to people. Tribal, traditional religions dominated this area, many involving the ritualistic drinking of blood from chimpanzees. At some time in the past, someone drank the blood from an infected chimpanzee and it then began to spread among humans through physical exchange of bodily fluids. HIV/AIDS became an epidemic disease through truck drivers spreading the disease along the main Congo highway. The drivers would often stop at a bar/hotel and partake in intravenous drug activity and sexual activity with prostitutes. Over time, people from equatorial Africa migrated to the Caribbean and then to the U.S. HIV/AIDS then became a pandemic disease.

Though HIV/AIDS originated in the Congo basin, today Southern Africa has the highest rates of the disease.

Language and Religion

The **Niger-Congo language family** dominates in sub-Saharan Africa, with the **Bantu language group** being the largest and most extensive Niger-Congo language subfamily.

The **Khoisan language family** dominates in southwestern Africa. This language family at one time covered most of central and southern Africa. Today, only several thousand people, mainly Bushmen, speak it. The Khoisan language is most distinctive in its sound, a distinctive "clicking type" sound.

Christianity prevails over most of sub-Saharan Africa, intermixed with a variety of traditional, tribal religions.

Sub-Saharan Africa Subregions

Sub-Saharan Africa consists of five subregions: Southern Africa, East Africa, Equatorial Africa, West Africa, and the Sahel.

Southern Africa

Southern Africa is the region's most economically developed area. Extensive and diverse minerals and agricultural crops, and relatively significant European settlement, are major reasons for the higher level of development. Southern Africa includes countries such as Angola, Namibia, Zambia, Zimbabwe, Mozambique, and South Africa.

Angola is a former Portuguese colony which has recently become an important oil producing country in sub-Saharan Africa.

Namibia is a former protectorate of South Africa. It includes most of the **Namib Desert**, which is the basis for the country's name. Namibia also includes most of the Bushmen who speak the Khoisan language.

Zambia and **Zimbabwe** are former British colonies originally known as Northern and Southern Rhodesia. Zimbabwe especially has encountered much political and ethnic instability leading to major economic and political turmoil.

Mozambique is a former Portuguese colony with extensive deposits of bauxite, the ore from which aluminum comes.

South Africa is Southern Africa's most populated country. It contains a wide array of important minerals, including coal, iron, uranium, gold, and diamonds. South Africa is the region's leading country in income and development. Historically, two main European groups settled in South Africa: the Dutch and the British. The Dutch predominated in the interior and the British became concentrated in the south around Cape Town. The descendants of the original Dutch settlers are known as the **Boers or Afrikaners**. The Afrikaners assumed political control of the country in the mid-1900s and instituted the apartheid policy. **Apartheid** was a policy of racial separation, especially subjecting the African population to widespread discrimination. Africans were forced to live in isolated, remote areas known as homelands. Over time, two of these former homelands became separate countries: Lesotho and Swaziland. South Africa is a classic multinational, multiethnic state. There are four main ethnic/racial groups: Africans, Coloreds, Europeans, and South Asians. Africans comprise about 80 percent of the country's population. The Africans include a wide variety of different tribal and language groups, including the Zulu and the Xhosa. The Coloreds are a mixed group of Africans and Whites, concentrated in the Cape Town area. The Europeans comprise less than 10 percent of the population, divided between the Afrikaners and the British. The smallest group is the South Asian population, descended from people from South and Southeast Asia. They were brought in by the Dutch and the British as domestics and laborers.

East Africa

East Africa includes much of the Great Rift Valley landscape. Most of the subregion is in the tropics but includes many highland areas that were attractive to European settlement. East Africa includes countries such as Kenya, Uganda, and Rwanda.

Kenya is the dominant country in East Africa. It is the most populated country and contains relatively large numbers of European settlers. Nairobi is the capital of Kenya and the largest city in East Africa. Mombasa is Kenya's main port and the busiest port in all of East Africa. Kenya also contains the greatest concentration of big game reserves. Safaris and animal hunting are important components of the economy.

Uganda, like Kenya, is a former British colony. It was the country formerly led by a brutal dictator, Idi Amin, who was the basis for the movie, *The Last King of Scotland*. Uganda has had much success in using public health education programs aimed at young people in reducing HIV/AIDS.

Rwanda is a small country east of the Congo and south of Uganda. Two main tribal/ethnic groups comprise the population of Rwanda: the Hutu and the Tutsi. The **Hutu** are the majority group in Rwanda. They are primarily farmers who are relatively short and squat in physique. The **Tutsi** are the minority group. They are mainly pastoralists (animal grazers) and are tall and more sophisticated in appearance. Under the Belgian rule, the Tutsis were a privileged minority. Over time the Hutus became jealous of the privileges awarded the Tutsis and ridiculed them as "cockroaches." When Rwanda became independent, the government gave each citizen two items: a transistor radio and a machete. The transistor radios were intended to connect the citizens to the rest of the world while the machete was meant to clear the tropical forests for crop or animal use. In April of 1994, the Hutu president attempted to fly to the neighboring country of Burundi. The plane carrying the president was shot down by a missile, killing the Hutu leader. Rumors quickly circulated, especially through Hutu radio disc jockeys, that Tutsis were responsible. The disc jockeys then encouraged the Hutus to go home, get their machetes, and kill Tutsis in revenge. Ethnic cleansing then ensued against Tutsis, young and old, male and female, lay and clergy. Up to a million people died with millions more becoming refugees.

Equatorial Africa

Equatorial Africa is the least economically developed subregion in sub-Saharan Africa. It is based on the Congo River basin and adjacent areas of central Africa. It has the highest concentration of endemic diseases in sub-Saharan Africa, poor soils, dense barrier forests, and much political and ethnic instability. The Ebola virus originated in this area as well as HIV/AIDS.

West Africa

West Africa is the most populated subregion in sub-Saharan Africa. Historically, it is the homeland of the Bantu, site of very early agriculture, urbanization, and political empires. West Africa includes countries such as Sierra Leone, Liberia, Ivory Coast, Ghana, and Nigeria.

Sierra Leone is a small country on the west coast of Africa. It was created to be a homeland for freed former British slaves. It is known for its "blood diamonds." Recently the Ebola virus greatly affected Sierra Leone. Much ethnic and political instability has also negatively affected the country.

Liberia is another small country on the west coast of Africa. It was created to be a homeland for freed former American slaves. The recent Ebola virus tremendously affected Liberia.

The **Ivory Coast (Cote d'Ivoire)** is located east of Liberia. It is a former French colony. In the 1980s, a Catholic president constructed a massive, hugely expensive basilica to rival St. Peter's in Vatican City. The basilica was built in the president's home town of Yamoussoukro in the interior, which became a forward capital, relocated from the former capital of Abidjan on the coast.

Ghana, once known as the Gold Coast, became the first West African state to achieve independence after World War II.

Nigeria is the most populated country in all of Africa. It is the region's leading oil producer. The original capital was Lagos, on the coast. **Abuja**, in the interior, is a more recent capital, an excellent example of a forward capital. Nigeria is another well-known illustration of a multiethnic, multinational state. There are three main ethnic groups in Nigeria: the Hausa, the Yoruba, and the Ibo. The **Hausa** dominate in the north; the **Yoruba** in the southwest; the **Ibo** in the southeast.

The Sahel

The Sahel is a transitional area extending west to east along the southern margins of the Sahara. In the north there is a dominance of Islam, Arabic languages, dry climates, animal grazing, and lighter-skinned people. In the south, there is a dominance of Christianity, Bantu languages, humid climates, agriculture, and darker-skinned people. It is a very fractious, turbulent area with lots of ethnic, religious, and political instability. It is also one of the world's most overpopulated and rapidly growing areas with great poverty and lots of environmental issues.

NORTH AFRICA AND SOUTHWEST ASIA

Where Is North Africa and Southwest Asia?

The region of **North Africa and Southwest Asia** includes the Mediterranean coast of Africa from Morocco to Egypt and the Sahara of North Africa as well as Turkey, the Middle East, the Arabian Peninsula, Iran, Afghanistan, and the central Asian countries formerly known as Turkestan.

Why Include North Africa and Southwest Asia in the Same Region?

There are a number of commonalities that link this vast global area. Dry climate conditions with low amounts of precipitation dominate most of the region. The Afro-Asiatic language family dominates in North Africa and much of the Middle East and Arabian Peninsula. Islam is the dominant religion of most of the region. Due to dry climates, most of the region is relatively less populated than sub-Saharan Africa. Historically, many important cultural developments originated here.

Major Qualities of North Africa and Southwest Asia

Spatially, this region is at a world crossroads interconnecting Europe, Africa, and Asia. Physically, dry or arid climate conditions dominate this area. It includes the world's largest known deposits and reserves of oil.

All three of the world's western and monotheistic religions originated here and spread to other places. Many early culture hearths, such as Mesopotamia and the Nile River Valley, began here.

Demographically, population is concentrated in discontinuous clusters around infrequent water sources. Today, there is much cultural and political conflict that affects peoples and states within the region.

Importance as a Culture Hearth

Two main early world civilizations began in North Africa and Southwest Asia. The Nile River Valley was the site of early cultural developments and kingdoms, including the building of the pyramids. The **Nile River** is the longest river in the world in terms of its length, flowing south to north from the mountains of eastern Africa to its end at the Mediterranean Sea. In Southwest Asia, **The Fertile Crescent** is an area stretching from the eastern Mediterranean coast to the head of the Persian Gulf. Within The Fertile Crescent there is a smaller area located between the Tigris and the Euphrates rivers. This smaller area is historically known as **Mesopotamia**, where the earliest archaeological evidence of cities and agriculture originated. Today, the modern state of Iraq comprises most of Mesopotamia.

Patterns of Religion

Three major world religions originated in Southwest Asia: Judaism, Christianity, and Islam. These three religions are all described as western and monotheistic in nature.

Judaism is the oldest of the western and monotheistic religions, tracing its ancestry to the prophet Abraham. It is numerically the smallest of the major world religions, with approximately 15 million adherents. Judaism only dominates in one country, Israel. Due to its small size and localized dominance, it is known as an **ethnic religion**. For Judaism, there is one major sacred site, the **Western Wall**, in the old city of Jerusalem. Jews believe this is the only remaining part of a temple largely destroyed by the Romans. Over time, three different philosophical branches of Judaism evolved. The most traditional branch is the **Orthodox branch.** The most liberal branch is the **Reform branch**. The **Conservative branch** is philosophically between the other two branches.

Christianity is a universalizing religion that grew out of the older and ethnic Judaism about 2,000 years ago. Jesus Christ, the founder of Christianity, was a Jew. Early on, Christianity spread from the eastern Mediterranean into southern Europe and became associated with European culture and globally spread with the extensive migrations of Europeans. The three branches of Christianity were discussed in the Europe region chapter.

Islam is the youngest of the western and monotheistic religions. It originated in Southwest Asia, in the Arabian Peninsula. Similar to Christianity, Islam is a universalizing religion. The founder of Islam is the prophet, **Muhammad**, who officially began the new religion in 622 AD in the city of Madinah. Islam shares many of the same prophets as Judaism and Christianity, including Abraham, Moses, and Jesus. The main holy book of Islam is the **Quran**, written in Arabic. Muslims believe the Quran is a book of revelations received by Muhammad from Allah through the angel Gabriel.

The essence of Islam is identified by the **Five Pillars of Faith**. The first pillar of faith is the reciting of the basic creed—"there is no God but Allah and Muhammad is his prophet." A second pillar of faith is daily prayer facing Makkah. A third pillar of faith is fasting from sunrise to sunset during the Muslim month of Ramadan. Almsgiving to the poor and needy is a fourth pillar. The fifth pillar is making a "hajj," a pilgrimage to Makkah during one's life.

There are three major sacred sites for Islam: Makkah, Madinah, and Jerusalem. **Makkah** (Mecca) is the #1 sacred site. This is the birthplace of the prophet Muhammad. It is also the location of the Grand Mosque, one of the main destinations of the hajj. **Madinah** (Medina) is the #2 sacred site. It is the city where the religion was founded by Muhammad and it is where the tomb of the prophet is found. **Jerusalem** is the #3 sacred site for Islam. Above the Jewish Western Wall is the **Temple Mount**. The Dome of the Rock and the Al Aqsa Mosque are both found on the site. The Dome of the Rock commemorates the ascent of Muhammad to heaven and back to earth.

Over time, two major branches of Islam emerged: the Sunni and the Shiite branches. There are differences between the two branches in terms of numbers/proportions, leadership, and philosophy. The **Sunni branch** is the majority branch of Islam. About 80 percent of all Muslims in the world are Sunni Muslims. Sunnis are the most widespread branch, dominating most of the Muslim world. The **Shiite branch** is the minority branch, dominating only in Iran and Iraq. Sunnis believe in an elected leadership whereas Shiites believe leadership must be traced directly back to the prophet Muhammad through his son-in-law Ali and grandson Hussein. Philosophically, Sunnis tend to be more moderate and accepting of outside ideas. Shiites tend to be more fundamental and traditional in their beliefs. The Shiites have two sacred sites of their own: Najaf and Karbala, both in Iraq. **Najaf** is the location of the tomb of Ali; **Karbala** is the location of the tomb of Hussein.

There are historical, philosophy, and language connections between Jews and Muslims. Historically, both Jews and Muslims trace their ultimate ancestry to the prophet Abraham. Jews are linked to Abraham through his marriage to Sarah and their eventual son Isaac. Muslims are linked to Abraham through his relations with a servant Hagar and their son Ishmael. Philosophically, both Jews and Muslims are monotheists. Linguistically, Jews and Muslims both speak languages, Hebrew and Arabic, which are part of the Afro-Asiatic language family and the Semitic language subfamily.

North Africa and Southwest Asia Subregions

North Africa and Southwest Asia consists of six subregions: The Maghreb and Libya, The Nile River Valley, The Arabian Peninsula, The Middle East, The Empire States, and Central Asia.

The Maghreb and Libya

The Maghreb refers to the northwest coastal states of Africa, and we also include the adjacent country of Libya. The Maghreb coast has more humid conditions than the rest of North Africa and supports a relatively dense concentration of people.

Morocco is located on the northwest coast of Africa, south of Gibraltar. It is a conservative kingdom that formerly was a French colony. It includes the cities of Casablanca, Marrakech, and Tangier.

Algeria is another former French colony that fought a bitter conflict for liberation in the 1950s and early 1960s. Many Algerians emigrated to France, where they form one of Europe's largest Muslim populations.

Tunisia is the smallest of the Maghreb states. In 2011, Tunisia was the origin of the Arab Spring. The **Arab Spring** was a wave of internal revolt, greatly assisted by the use of social media, against repressive and corrupt governments in the Arab World. The Arab Spring eventually spread to Libya and Egypt and finally to Syria.

Libya is not officially a part of The Maghreb, but is often included in the same subregion. Libya is an important oil producer and exporter. During the Arab Spring of 2011, the dictator Qadhafi was overthrown and killed.

The Nile River Valley

The Nile River Valley includes the basin of the Nile River, the longest river in the world.

Egypt is considered to be the **heart of the region**. Its centralized position west to east, its prominence as a historic early world culture hearth, its being the most populated Arab country, and its influence within Islam all combine to make Egypt the center of the region. The Arab Spring in Egypt led to the overthrow of President Mubarak.

Sudan is a largely desert country centered on the confluence of the Blue Nile and White Nile. Decades of civil war between north and south impoverished the country and eventually led to the political separation of Sudan and South Sudan. A particularly violent part of Sudan's recent history occurred in the western province of Darfur. In **Darfur**, hundreds of thousands of people were killed and several million people were driven from their homes.

The Arabian Peninsula

The Arabian Peninsula lies south of Jordan and Iraq, adjacent to the Red Sea, the Arabian Sea, and the Persian Gulf.

Saudi Arabia occupies most of the land area of the peninsula and has the largest population. Saudi Arabia is the world leader in oil production and has the largest world concentration of oil reserves. The country includes the two most important sacred places in Islam, Makkah and Madinah.

Kuwait is an oil-rich country at the head of the Persian Gulf. Iraq's invasion of Kuwait in the early 1990s led to the First Gulf War.

Bahrain is an island state ruled by a royal family. It is a major banking center, with a predominantly Shiite population. Nearly two-thirds of its work force is foreign. The U.S. Navy's Fifth Fleet is headquartered here.

Qatar is a peninsula country neighboring Bahrain. It is rich in oil and natural gas. Qatar is home to the media giant Al Jazeera.

The **United Arab Emirates** is a federation of various emirates. The two most well-known emirates are Abu Dhabi and Dubai. These two emirates have the area's largest reserves of oil. Dubai includes the Burj Khalifa, the world's tallest building.

Yemen occupies the southwest part of the Arabian Peninsula. A Shiite rebellion against the central government has destabilized the north and a secessionist movement occurs in the south. Al-Qaeda and other terrorist groups are highly concentrated within the country.

The Middle East

Five countries comprise the Middle East: Iraq, Syria, Jordan, Lebanon, and Israel.

Iraq is the largest state in population and area within the subregion. Iraq is a multinational state, comprised of three main ethnic groups: Shiites, Sunnis, and Kurds. The **Shiites** are the overall majority group, dominant in the south. Under the rule of Saddam Hussein, the majority Shiites were greatly discriminated against. The **Sunnis** are dominant in the central and western areas. They controlled the government under Hussein, who was overthrown in the Second Gulf war. The **Kurds** are dominant in the north. Similar to the Shiites, the Kurds were greatly discriminated against under the Hussein administration. As mentioned previously, the Kurds are a classic example of a stateless nation. Complicating things even more, the terrorist group **ISIS (the Islamic State of Iraq and Syria)** is involved in much of the conflict in Iraq.

Syria is a multinational state ruled by an oppressive leader, Bashar al-Assad. The Arab Spring was unsuccessful in Syria in overthrowing the current administration. There is tremendous internal instability in Syria. Government forces under Assad are supported by Russia; anti-government forces are aided by the U.S.; and **ISIS** is headquartered in Syria.

Jordan is a moderate-sized, poor country, lacking oil resources. The majority of the population is ethnic Palestinian.

Lebanon is a small country on the eastern Mediterranean coast. It contains many religious and ethnic factions that create much political instability. An Iran-sponsored terrorist group, **Hizbollah**, is a major political force in southern Lebanon. Hizbollah has been involved in many anti-Israeli activities and conflicts.

Israel is a Jewish state created in 1948. That year, the British pulled out of their Palestine territory. Israel is distinctive in being neither Arab nor Muslim in dominance. Soon after proclaiming Israel as a state, Egyptian, Iraqi, Jordanian, and Syrian forces invaded in the 1948 War of Independence. In 1967, the Six-Day War resulted in another Israeli victory and Israel gained control of the Sinai Peninsula, Gaza Strip, West Bank of the Jordan River, and the Golan Heights on the Syria border.

Eventually, the Sinai Peninsula was given back to Egypt. The Golan Heights, the Gaza Strip, and the West Bank remain as **Israeli occupied territories**. The Palestinian Arabs, like the Kurds, are another example of a stateless nation in this region. In the Gaza Strip, a radical Palestinian group, Hamas, is dominant.

The Empire States

The **Empire States** consist of two countries, Turkey and Iran. They are known as "empire states" due to their history. Turkey was the historical basis for the Ottoman Empire and Iran was the basis for the Persian Empire. They are both demographically significant as they are the most populated countries in the whole region. Culturally, the Turks and the Iranians do not speak Arabic languages; the Turks speak an Altaic language and Iran speaks an Indo-European language. Religiously, Iran is dominated by Shiite Islam and Turkey is essentially a secular state.

The Ottoman Empire collapsed after the end of the Second World War. A charismatic Turk named Ataturk created the **modern state of Turkey**. He proclaimed Turkey to be a secular state where Islam would play a minor role in the country's affairs. He also moved the capital from Istanbul to Ankara. Ankara is a classic example of a **forward capital**. Ataturk thought that Istanbul, on the coast, was too prone to outside attack and a more inward location would be militarily more suitable. Also, Istanbul is technically on the European mainland whereas most of Turkey is in Asia. Ataturk felt that symbolically Istanbul was too closely connected to former European influences. Ankara is in the Anatolian Plateau, the root of basic Turkish culture. Ankara is also in a more centralized location and could act as a unifier rather than more remote Istanbul.

Iran is an oil-rich state located between the Caspian Sea and the Persian Gulf. Iran has the second largest concentration of oil reserves in the region. Whereas Turkey is a secular state, Iran is a theocracy. A **theocracy** refers to a place where government and religion are intertwined very closely. The leading political figure in Iran is the leading Shiite cleric. Iran is strongly anti-Israel and has significant aspirations to be a nuclear power. Iran has been closely involved in a lot of the turmoil and conflict within the region.

Central Asia

Central Asia includes the historic area known as Turkestan as well as the adjacent country of Afghanistan.

Turkestan refers to the five former Soviet-controlled republics that are now politically independent. These five states are dominantly Islamic in religion and most people speak a Turkish-related language. Both cultural elements are consequences of the Ottoman Empire. Each of the Turkestan states is dominated by different ethnic and language groups.

Kazakhstan is the largest Turkestan state in land area. The Kazakh ethic group is the major ethnicity. It is economically dominated by cotton, uranium, and oil production. It is the nuclear waste depository for the former Soviet Union. Soviet and Russian missiles are launched and returned to Kazakhstan.

Uzbekistan is the most populated Turkestan state. It is centrally located within the area and adjoins all the other states. The Uzkbeks are the dominant ethnic group.

The three other Turkestan states are **Turkmenistan** (dominated by the Turkmen), **Kyrgyzstan** (dominated by the Kyrgyz), and **Tajikistan** (dominated by the Tajiks).

Afghanistan is the non-Turkestan country in Central Asia. Afghanistan is dominated by Indo-European speaking peoples and was not a part of the Soviet Union. It is a multinational state. The largest ethnic group is the **Pushtuns** (Pashtuns), who dominate in the south. The Tajiks are the second largest group, dominating in the northeast. The Hazaras dominate in the central area, west of Kabul, the capital. The Uzbeks are the fourth largest group, dominating in the north.

Up until the 1970s, Afghanistan was a monarchy. In the late 1970s, a Soviet-supported military revolutionary council seized power and created a Marxist government. To support the Marxist council, Soviet troops invaded in 1979. The Muslim opposition, the **mujahideen**, became supported by the U.S. and the Saudis. In 1989, the Soviets were forced to withdraw. In the power vacuum created by the Soviet withdrawal, a group of young militant Muslim religious students, the **Taliban**, associated with the Pushtun ethnic group, emerged. This group imposed an extreme interpretation of Islamic law, forcing women to not be formally educated, to be subservient to males, and to stay in the house as obedient wives. Other Afghan groups began to oppose the harsh rule of the Taliban. Derived from the Taliban, another radical group emerged, **Al-Qaeda**. Al-Qaeda was led by a former Saudi citizen who fought with the mujaheedin against the Soviets, Osama bin Laden. It was Al-Qaeda who claimed responsibility for the 9/11 attack on U.S. soil. In retaliation, the U.S. and Britain invaded Afghanistan and removed the Taliban from political control. Events in Iraq distracted the U.S. and the Taliban regrouped into the mountains, became more sophisticated with weaponry, and reemerged as a power force in Afghanistan.

SOUTH ASIA

Where Is South Asia?

South Asia refers to a region located in the south central part of Asia. Mountains and deserts physically separate it from the rest of the Asian continent. India is the dominant country in area size and population within the region. Given the size of the territory and the regional dominance of India, this region is often referred to as the **Indian subcontinent**. The countries of Pakistan, India, Bangladesh, Nepal, Bhutan, and Sri Lanka comprise South Asia.

Major Qualities of South Asia

Currently, South Asia is the most populated world region. It contains about 25 percent of the world population. India is the #2 populated country in the world, only exceeded by China. Not only is there a large population in South Asia, but it is growing quite rapidly, raising concerns about whether food production can keep pace with the population growth.

South Asia is one of the economically poorest regions in the world. Income levels are low, poverty is widespread, and standard of living is low. Given the low level of development, subsistence agriculture is the dominant economic activity.

The two dominant countries in the region, India and Pakistan, have been involved in numerous conflicts since their independence in 1947. Both countries today have nuclear weapons.

The Monsoon

South Asia's weather and climate are dominated by a monsoon climate. A **monsoon** is defined as a seasonal change in wind direction. It is a phenomenon most evident and dominant in South Asia, where there is a large expanse of land that is adjacent to large bodies of water. Land and water heat and cool differently from season to season.

In the summer in South Asia, the Asian landmass quickly and intensely heats up. The large mass of warm air, being light, rises and forms an intense low pressure system. At the same time, the Indian Ocean to the south is cooler than the land. The cool air, being heavy and dense, sinks and forms a high pressure system. Wind always flows from high pressure toward low pressure. In the summer, the dominant flow is from the southwest toward the northeast. The **summer southwest monsoon** brings in lots of moisture off the water onto the land. Mountains intensify the effect, causing much rainfall, especially in the northeast.

In the winter, conditions are reversed. The Asian landmass quickly and intensely cools off, forming high pressure. The Indian Ocean is warmer in the winter, forming low pressure. Thus, a reverse wind flow occurs. In the **winter northeast monsoon**, dry conditions result over South Asia.

Landform Patterns

There are four main landform areas in South Asia. The **Himalayan Mountains**, and adjacent mountains, dominate in the north. South of the mountains are the river lowlands. In Pakistan, the **Indus River** dominates. In north India, the **Ganges River** is the dominant river that connects with the **Brahmaputra River** from the northeast to form a large delta in Bangladesh. In central India, the **Central Indian Plateau** is a transition area of hills and rolling topography. The **Deccan Plateau** occupies most of southern India. It is a volcanic tableland that descends to the narrow coastal plains below.

Culture History

The **Dravidians** are the native occupants of South Asia. They are darker-skinned peoples who today dominate in the Deccan Plateau. They speak the Dravidian language, which is a different language family from most of the rest of South Asia.

The **Indo-Aryans** are lighter-skinned peoples who came from the northwest around 1500 BC. They introduced the Indo-European languages and the basis for Hinduism into South Asia.

Language Patterns

Indo-European languages predominate in northern and western parts of South Asia.

Indo-European languages include Hindi (one of the official languages of India), Punjabi (on the border of India and Pakistan), Urdu (a traditional language of Pakistan), and Bengali (along the India and Bangladesh border).

Dravidian languages dominate in the Deccan Plateau. Tamil is a major language in southeast India and northern Sri Lanka.

Religious Patterns

Hinduism predominates in India and Nepal. **Buddhism** is dominant in Sri Lanka and is the state religion in Bhutan. **Islam** is the dominant religion in Pakistan and Bangladesh. **Sikhism** predominates in the Punjab area of western India.

Hinduism is considered by many scholars as the oldest major world religion. It is not just a religion, but a culture, society, and religion interconnected. It is a complete way of life for its adherents. There is no common creed, no single doctrine, and no identifiable single founder. Hinduism is an **ethnic, polytheistic, and eastern** religion. Three main ideas are the essence of the Hindu religion. **Karma** refers to the Hindu belief that a person's soul or spirit can be transferred after the death of the body. Karma also refers to the Hindu belief that the past affects the present which affects the future. **Reincarnation** is the Hindu belief that, at death, the soul or spirit can be reborn in another form. The **caste system** is a rigid hierarchy of social classes that is a traditional component of Hindu belief. The Ganges River is a major sacred space for Hinduism.

Buddhism is a major world religion that, like Hinduism, originated in the Ganges River area of northern India. Siddhartha Gautama is the founder of Buddhism. He was born a Hindu prince, later rejected his princely status, and founded a new religion known as Buddhism. From its beginnings, Buddhism rejected the caste system. Unlike Hinduism, Buddhism is the third main global universalizing religion. The essence of Buddhism is the **Four Noble Truths**. The first truth is the Buddhist belief that life involves sorrow and suffering. The second truth is that the cause of sorrow and suffering is human desire. People suffer because they want things they cannot have. The third truth is that one can escape suffering by ending desire, stop wanting, and reach a stage of not wanting. The fourth truth is the belief that to end desire, you follow a middle path, avoiding extremes of too much pleasure and desire and too much asceticism. This middle path includes meditation and contemplation. Two main spatial branches of Buddhism have evolved. **Mahayana Buddhism** spread northward to China, Korea, and Japan. **Theravada Buddhism** spread south and east to Sri Lanka and Southeast Asia.

Islam originated in Southwest Asia but a group of Muslims, the **Mughals**, arrived in South Asia around 700 AD and introduced Islam. For about 1,000 years, they controlled most of this region. Their greatest landscape impact is the Taj Mahal, built by a Mughal prince as a tomb for his deceased wife. Today, Islam dominates in Pakistan and Bangladesh and forms the largest religious minority in India. Islam and Hinduism are largely incompatible. Islam is monotheistic, universalizing, does not condone idols, has one sacred book (the Quran), a uniform dogma (The Five Pillars of Faith), eats beef, buries their dead, and theoretically teaches social equality. In contrast, Hinduism is polytheistic, ethnic, reveres many idols, has various sacred writings, varying beliefs, venerates cows, cremates the dead, and includes the rigid caste system.

Sikhism is a hybrid, blended religion that originated in and dominates the western Indian state of Punjab. The Punjab is a transitional area that lies astride the India/Pakistan border. Pakistan is dry and mountainous whereas India is humid and more lowland in nature. Islam dominates Pakistan and Hinduism dominates India. In the 1400s, Sikhism arose in the Punjab to try to bring together these two major disparate religions. Sikhism incorporates elements of Hinduism and Islam. Like Hinduism, Sikhs believe in reincarnation. Like Islam, Sikhs are monotheists. The most sacred space for Sikhism is the Golden Temple at Amritsar, the capital of Punjab.

Geopolitical Framework

The Mughal Empire controlled most of South Asia up to the 1700s. By then, the empire had weakened and lots of political and ethnic tensions had emerged. In the power vacuum, a private British company, The **British East India Company**, assumed political and economic control. In the mid-1800s, a mutiny against the East India Company directly brought in British troops and the British government. During British control, Hindu, Muslim, and Sikh rulers retained much local authority. By the 1920s, political protests began to call for eventual independence. In the 1930s, a charismatic Hindu, **Mahatma Gandhi**, used a campaign of nonviolence to galvanize much of the opposition to the British.

After the end of World War II, in 1947, the British finally decided to leave their Indian empire. An agreement called for two new states to emerge, Pakistan and India. How and where to draw the boundary separating the two new states? It was agreed that the existing pattern of **religion** would be the basis for the split. Areas dominated by Hindus would become India. Areas dominated by Muslims would become Pakistan. At the time, Muslims were concentrated in two separate places, to the west and to the east of India. Thus, Pakistan began as a fragmented state, with two parts: West Pakistan and East Pakistan. The capital was in West Pakistan. India would not allow any movement between the two parts of Pakistan across India. Increasingly, people in East Pakistan became isolated from West Pakistan. In the early 1970s, **East Pakistan seceded from Pakistan and became the independent country of Bangladesh**. West Pakistan became the Republic of Pakistan.

South Asia Subregions

South Asia is comprised of the subregions of Pakistan, India, Bangladesh, Nepal, Bhutan, and Sri Lanka.

Pakistan

Pakistan is an Islamic republic, overwhelmingly dominated by Muslims. About 80 percent of the population is Sunni Muslim, with Shiite Muslims comprising most of the rest of the population. Originally, Pakistan included present-day Bangladesh, which became independent in the early 1970s. The traditional capital was the large port city of Karachi. Relatively recently, the capital was relocated to a **new forward capital, Islamabad**. Islamabad is located in the northern interior, close to an embattled transitional area along the Pakistan/India/ China border known as **Kashmir**.

When partition occurred in 1947, existing states were given the choice of being incorporated into India or Pakistan. In Kashmir, the great majority of the population was Muslim but the local ruler was Hindu. When he decided not to join Pakistan, a Muslim uprising, with Pakistan's support, occurred. The local ruler then called for help from India. Eventually a ceasefire line was drawn, with most of the territory in India's control. Pakistan has always demanded a referendum so that people themselves could decide to be in India or Pakistan. India has refused such a referendum. Both countries have been in conflict over Kashmir several times since 1947. Complicating factors today are that both India and Pakistan have nuclear capabilities, and China also claims part of the Kashmir area.

Much of the northwest area of Pakistan is controlled by the Taliban who give aid and support to Al-Qaeda. This was the part of Pakistan where Osama bin Laden was captured and killed.

India

India is the dominant country in South Asia, with over a billion people and encompassing about 75 percent of the total area. It is currently the second largest country in the world. By 2030, it will likely surpass China to become the most populated world state. It is a political democracy; in fact, India is the most populous democracy in the world.

Mumbai, on the west central coast, is India's largest industrial and financial city. It is also home to "Bollywood," India's equivalent of Hollywood. The Indian entertainment industry is a major element of the city's economy. Mumbai is the dominant city in India's west.

New Delhi is India's capital city, located within the vast Ganges River Valley in the north of the country. The Delhi/New Delhi area is India's largest urban concentration.

Varanasi (Benares) is India's holiest city and one of the world's oldest living cities. It is located along the north bank of the Ganges River. There are numerous temples devoted to various Hindu gods along the Ganges in Varanasi.

Bangalore, in the southern Deccan Plateau, is the center of India's high-tech computer industry.

Kolkata (Calcutta) is India's dominant city in the east.

Bangladesh

Bangladesh occupies the delta of the combined Ganges and Brahmaputra Rivers. It was born as an independent country in 1971, following secession from Pakistan. It is extremely vulnerable to environmental hazards, including flooding and cyclone (hurricane-like) storms. Bangladesh is one of the most densely populated countries in the world.

Nepal

Nepal is a buffer state between China to the north and India to the south. It is predominantly Hindu in religion and Hinduism is the country's official religion. Nepal is dominated by the Himalaya Mountains.

Bhutan

Bhutan is another mountainous country wedged between India and China's Tibet area. Buddhism is the official religion of Bhutan and it includes the majority of the population. Like Nepal, Bhutan is a poorly developed economy. The national goal of Bhutan is happiness.

Sri Lanka

Sri Lanka is an island state off the southeast coast of India. There are two main ethnic groups that comprise the population of Sri Lanka: the Sinhalese and the Tamils.

The **Sinhalese** are the majority group. Their ancestry lies in northwestern India. They speak an Indo-European language, Sinhala. They are religiously Buddhist.

The **Tamils** are the minority group. They came from southern India. They speak Tamil, a Dravidian language. They are religiously Hindu. They are concentrated in the northern and northeastern parts of the island.

A civil war occurred between these two groups, starting in the mid-1980s. Traditionally, the Buddhist Sinhalese discriminated against the minority Tamils. The majority Sinhalese government finally defeated the radical Tamil Tigers around 2009.

EAST ASIA

Where Is East Asia?

East Asia refers to a region in the eastern part of Asia. East Asia includes the countries of China, Mongolia, North Korea, South Korea, Taiwan, and Japan.

Major Qualities of East Asia

East Asia is the second most populous world region. It contains about 25 percent of the world population. Only South Asia has more people. China is the most populated country in the world, with about 1.4 billion people.

There is a very wide spectrum of level of economic development in East Asia. Japan, South Korea, and Taiwan are all highly developed economies. China is one of the fastest growing economies, though still classed as less developed. Mongolia is a relatively poor, less developed country. North Korea is one of the economically poorest countries in the world. East Asia is a major part of today's world economy and an emerging center of political power.

Our major emphasis will be on China, with a secondary focus on Japan.

Eastern versus Western China

Eastern and Western China differ in terms of topography, climate, and population. Eastern China is mainly comprised of river and coastal lowlands, while Western China is mainly mountains and high plateaus. Eastern China is primarily humid whereas Western China is mainly dry. Eastern China is densely populated while Western China is very sparely populated.

China versus United States

We will compare and contrast China and the United States in terms of climate, topography, area size, and population.

Both countries are quite similar in climate patterns. China and the United States have humid climates in the east and dry climates in the west. China and the United States have mainly river and coastal lowlands in the east and mountains and high plateaus in the west. The overall land area size is almost identical between the two countries. Both countries have dense populations in the east and more sparse populations in the west. A significant difference is population size. China is the most populated country in the world, with about 1.4 billion people. Although the United States is the third most populated country in the world, its population is about 315 million people.

Main Rivers of China

There are two main rivers in China: the Chang Jiang and the Huang He.

The **Chang Jiang** (also known as the Yangtze) is the principal river of south central China. It is China's longest river, and one of the longest rivers in the world. The Chang Jiang originates in the Tibetan Plateau and flows into the East China Sea. Shanghai, China's largest city, lies at the mouth of the Chang Jiang. The Chang Jiang is China's major river for movement of goods and people. The **Three Gorges Dam** lies in the central part of the Chang Jiang. It is the world's largest hydroelectric dam. It is a multipurpose project, intended to control flooding, generate electricity, and improve navigation further into China's interior.

The **Huang He** (also known as the Yellow River) is the main river of North China. The Huang He also originates in the Tibetan Plateau, makes a big inverted-U loop around the Ordos Desert, and then flows through the North China Plain before emptying into the Bohai Gulf. **The North China Plain** is the historic core area of Chinese culture, including the capital city of Beijing.

Chinese Culture History

For about 4,000 years, China's history was dominated by a series of dynasties. The most formative dynasty for Chinese culture is the **Han (Qin) dynasty**. It spanned the period of about 200 BC to 200 AD. Ethnic Chinese describe themselves as Han Chinese. The last dynasty is the **Manchu (Qing) dynasty**. It was the most extensive dynasty in area size and lasted from the mid-1600s to the early 1900s. In 1911, the last Chinese emperor was replaced during the Japanese invasion of that time.

Chinese Religious Complex

There are three main religious philosophies (the **Chinese religious complex**) that historically influenced China: Buddhism, Confucianism, and Taoism.

Buddhism originated in South Asia and spread extensively through East and Southeast Asia. When Buddhism diffused into China, it got intermixed with two native Chinese philosophies, Confucianism and Taoism.

Confucianism is a philosophy that dates from the sixth century BC. It was developed by an early Chinese philosopher, Confucius. Confucianism has had a significant influence on Chinese culture. Confucianism is a "behavioral philosophy." It instructs people on how to behave toward other people. Confucianism stresses obedience to authority, loyalty to the family, reverence to the elderly, importance of education, and following proper social behavior. It is the basis for the "Asian Way of doing business."

Taoism is a philosophy derived from the teachings of Lao-Tzu. It is an "ecological philosophy." It is rooted in nature worship and deals with how people should behave toward their environment. Taoism also involves mystical healing practices. The concepts of "yin" and "yang" derive from Taoism. **Feng shui,** an interior and architectural design philosophy, also derives from Taoism. Feng shui incorporates how best to place buildings on a site or how best to arrange furniture in a room to achieve proper harmony.

Twentieth Century Political Developments

By the 1940s, a full-scale civil war broke out in China between the Communists, led by **Mao Zedong**, and the nationalists, led by **Chang Kai-shek**. In 1949, the communists were victorious and the defeated nationalist fled to the offshore island of Taiwan. Mao formed the People's Republic of China, based on the communist political and economic philosophy.

In the late 1950s and early 1960s, Mao initiated **The Great Leap Forward**. It was an overly ambitious policy of trying to enforce a labor-intensive industrialization at the expense of agriculture. The result was grossly inferior products and a depressed agricultural system. Tens of millions of people died under a repressive and ruthless administration.

From the mid-1960s to the mid-1970s, Mao launched another brutally repressive strategy, **The Cultural Revolution**. It was a policy of attempting to ensure a long-term communist society by using radical young people, the **Red Guards**. Led by the Red Guards, attacks occurred against parents, lawyers, and teachers to rid China of elite bourgeois elements. People were instructed to dress alike and think alike. The economy suffered and food and industrial production decreased. Again, tens of millions of people died from torture and starvation.

After Mao's death in 1976, China's economy began to recover and eventually become greatly transformed into a rapidly growing economy geared toward exports with global impacts.

China's Population Issues

China has about 1.4 billion people, the most populated country in the world. By the mid-twentieth century China was one of the fastest growing countries, growing about 3 percent per year. The post-Mao Chinese government realized that China needed to slow down its population growth to ensure future economic growth.

The government adopted the **one-child policy**. By the late 1980s, growth had slowed to about 1.2 percent; today the growth is only about 0.5 percent. Although very successfully lowering population growth, the policy has had many negative, unintended consequences. In most Asian cultures, including China, preference has always been to favor boy births rather than girl births. With the one-child policy, Chinese preferred a male child rather than a female child. With the availability of sonograms, the gender of the fetus could be determined. If a female fetus was detected, there is much evidence of the abortion of the female fetus, hopefully allowing the opportunity to have another child, preferably male. If the female fetus was not aborted, evidence indicates a lot of female infanticide occurred. If neither abortion nor infanticide happened, many girls were placed in orphanages. The overall result is one of the most unbalanced gender ratios in the world (approximately 125 males per 100 females).

Very recently, the Chinese government re-evaluated the policy. The policy still persists but two children are now allowable for many citizens.

Subregions of East Asia

East Asia includes the subregions of China, Mongolia, North Korea, South Korea, Taiwan, and Japan.

China

The core area of China is **China Proper**. China Proper refers to the eastern part of China, which is climatically humid, topographically more lowlands, densely populated, ethnically Han Chinese, and includes most of the agriculture, industry, and large cities of China.

China today is one of the fastest growing economies of the world. The economic growth is uneven, though. Most economic benefits and successes have primarily occurred in the coastal regions and secondarily to Beijing. These areas are in China Proper. In contrast, China's interior and northeast portions have seen little economic expansion.

Most of the growth coincides with China's **Special Economic Zones**. The Special Economic Zones are concentrated along the east coast, especially around Shanghai and on the coast opposite Taiwan and Hong Kong, such as Shenzhen. The Beijing area is another growth core. In these Special Zones, the Chinese government highly encourages foreign investment.

China's **Rust Belt** is in the Northeast China Plain, historically known as Manchuria. Here cities are economically distressed, with many people and industries leaving for the south.

Mongolia

Mongolia is a sparsely populated, landlocked, and isolated state on the boundaries of China and Russia.

Mongolia is mainly semi-desert and desert; historically, it is the home of the Mongol Empire. Today, it contains a vast array of minerals, such as gold, copper, and coal. Chinese involvement and economic investment have been growing steadily.

North Korea

North Korea occupies the northern half of the Korean peninsula. After World War II, Communists expanded control into the north whereas non-Communists dominated in the south. From 1950–1953, the Korean War became an international event between communists and non-communists. A stalemate resulted in an armistice agreement that used the 38th parallel (line of 38 degrees North latitude) as a dividing line, now separating two independent countries: North Korea and South Korea.

North Korea is one of the economically poorest and most regimented countries in the world.

South Korea

South Korea is one of the most economically developed countries in the world today. Its capital, Seoul, is very modern and well-developed. From a dictatorship in the 1970s, South Korea has become a successful democracy.

South Korea is one of the so-called **Four Asian Tigers or Dragons**; Hong Kong, Singapore, and Taiwan are the other three Tigers. These four countries, in the 1960s and 1970s, made the rapid transformation from the less developed world to the more developed world.

Taiwan

Taiwan is an island off the east central coast of China. After the communists in China led by Mao Zedong defeated Chang Kai-shek in 1949, the nationalists fled to Taiwan.

Taiwan, like South Korea, is one of the Four Asian Tigers. Today, it is a modern, industrial, and well-developed state. China still refers to it as a breakaway province, eventually to be reunited with the mainland.

Japan

Japan is an island state east of the Korean Peninsula. Four main islands comprise Japan: Hokkaido, Honshu, Kyushu, and Shikoku.

Honshu is the largest island in area size and population. Tokyo, the capital and largest city, is on this island. Honshu also includes the **Tokaido megalopolis**, a collection of coalesced metropolitan areas that stretches from Tokyo on the northeast through Yokohama to Osaka, Kobe, and Kyoto on the southwest. Kyoto is the historic cultural center and former capital of Japan. Hiroshima, on the southwest coast, was one of the Japanese cities where the U.S. dropped an atomic bomb to help end World War II.

Hokkaido is the northernmost of the main Japanese islands. The **Ainu**, the indigenous people of Japan, are mainly concentrated today on Hokkaido.

Shikoku is the smallest of the main Japanese islands. **Kyushu** is the southernmost main Japanese island. It includes the city of Nagasaki, another city subjected to the dropping of an atomic bomb by the U.S.

Buddhism diffused to Japan from China and Korea. In Japan, Buddhism became intermixed with Shintoism. **Shintoism** is a native Japanese religious philosophy, closely bound to Japanese nationalism. It is a place- and nature-centered religion, including beliefs about harmony and balance with nature.

Japan did not emerge as a unified state until about the seventh century AD. Over time, Japan was divided several times. Until the mid-1800s, Japan was a medieval and feudal society. **Shoguns** were local and regional rulers who used warriors, the **samurai**, to enforce their rule. In the 1600s, the most powerful shogun, the Tokugawa shogun, reunited Japan and established Japan as an isolated state in the **Tokugawa Shogunate**. In the 1860s, the **Meiji Restoration** restored the emperor as the monarch of Japan, moved the capital to what is now Tokyo, and initiated a process of modernization and industrialization.

By the early 1900s, Japan was a major global leader in industry and military power. It expanded its control to the Korean Peninsula and invaded Manchuria in the 1930s. In 1941, Japan attacked Pearl Harbor. In 1945, the U.S. hastened the end of World War II by dropping the atomic bombs on Hiroshima and Nagasaki.

Although militarily defeated in World War II, Japan arose from physical destruction to become a global leader in industry and economic development by the 1970s and 1980s.

What factors created the **Japanese economic miracle**? Japan lacks major mineral resources, such as coal and oil, which usually lead to development. Instead of mineral resources, Japan capitalized on its highly homogeneous, intensely nationalistic population. Japanese workers also possessed a high work ethic; combined with historically low wages, economic production became very profitable. After the end of World War II the U.S. rewrote the Japanese constitution, promising to defend Japan if necessary. This allowed monies that would have gone to military use to be devoted to improving the economy. The long-term advantage of the war's physical destruction was that Japan rebuilt its economic infrastructure with modern machinery and technology. A huge advantage was the tremendous financial aid given to Japan by the U.S. and Europe.

Chapter 11

SOUTHEAST ASIA

Where Is Southeast Asia?

Southeast Asia is a region that occupies the southeastern part of Asia. Southeast Asia includes the mainland countries of Myanmar, Thailand, Laos, Cambodia, and Vietnam as well as island countries such as Malaysia, Singapore, Indonesia, and The Philippines.

Major Qualities of Southeast Asia

Southeast Asia is a physically fragmented region of numerous peninsulas and islands. It is also a culturally fragmented area with various ethnicities, languages, and religions. Similar to Eastern Europe, Southeast Asia exhibits the characteristics of a **shatter belt**, with internal and external factors causing a splintering and fracturing of the region. Historically, there are major impacts of European, American, Chinese, and Indian influences. Political instability and conflict are hallmarks of the region.

Southeast Asia is a secondary population cluster, with about 15 percent of the world's total people. Indonesia is the dominant country in terms of area and population, being the world's fourth most populated country.

The physical geography of the region is dominated by high relief, crustal instability (part of the Ring of Fire), and tropical climates.

Structurally, Southeast Asia includes a Mainland Subregion as well as an Insular Subregion.

Physical Geography of the Mainland

The Mainland Subregion consists of a series of rugged uplands interspersed with broad river lowlands. Four major rivers dominate the Mainland: the Irrawaddy, Chao Phraya, Mekong, and Red.

The **Irrawaddy River** flows through the country of Myanmar (Burma) in a wide valley with the largest city of Yangon in its broad delta. The **Chao Phraya River** flows through the country of Thailand, including the capital and largest city of Bangkok. The **Mekong River** originates in China, flows along the border of Thailand and Laos, then through the middle of Cambodia, including the capital Phnom Penh, and finally ends in a wide delta in southern Vietnam. The **Red River** also originates in China and flows through the lowland of northern Vietnam, including the capital of Hanoi.

Southeast Asia as a Shatter Belt

There are a large variety of different ethnic/language groups that internally fractured and shattered the region of Southeast Asia. These groups include the Thai, Tibetan-Burmese, Vietnamese, Mon-Khmer, Indonesian, and Chinese. Historically, this array of varied groups has created a complex ethnic mosaic.

In addition to a variety of ethnic and language groups in Southeast Asia, there is also much religious diversity. **Buddhism** dominates over most of Mainland Southeast Asia. **Confucianism** is highly present and influential in Vietnam. **Islam** predominates in Malaysia and Indonesia. **Roman Catholic Christianity** predominates in The Philippines.

External factors have also acted to shatter and fragment the region. The Mainland part of Southeast Asia is often referred to as "Indochina." **Indochina** refers to the vast influence of **South Asians** from the west and the **Chinese** from the north. The first major South Asian influence arrived about 2,000 years ago. Art, music, and religion diffused into Southeast Asia, including early impacts of Hinduism. The second wave of influence brought Buddhism in the thirteenth century AD. The Chinese influence was mainly through migration, especially in Vietnam.

A diversity of European political colonialization has also acted to shatter and fragment Southeast Asia. The **British** had colonies in Burma (now Myanmar), Malaya (now Malaysia), and Singapore. The **French** had a colonial empire including Laos, Cambodia, and Vietnam (French Indochina). The **Dutch** controlled the Dutch East Indies (now Indonesia). The **Spanish** controlled the Philippines until 1898, when the U.S. took control after the Spanish-American War ended. The Philippines was under **U.S.** political control from 1898–1946. **Thailand** (formerly Siam) was the only Southeast Asian country not under European/U.S. colonial control.

Territorial Shapes

Southeast Asia is a classic world area to investigate the different types of political territory that exist. Five main territorial shapes are commonly identified: compact, elongated, protruded, fragmented, and perforated.

A **compact state** has a fairly regular visual shape, such as circular or square or rectangular. Compact states have advantages of easier interconnections within the area and better opportunities to politically control the territory. In Southeast Asia, Cambodia is a classic example of such a compact state. Other global examples include Uruguay in South America, Zimbabwe in Africa, Hungary and Slovakia in Europe, and Mongolia in Asia.

An **elongated state** has a long, narrow shape. Elongated states have potential problems in effectively managing and controlling a vast elongated territory. Most elongated states extend north and south, often covering different climate areas over a latitudinal range. In Southeast Asia, Vietnam and Laos are classic examples. Other global examples include Chile in South America, Mozambique in Africa, Norway and Sweden and Finland and Italy in Europe.

A **protruded state** is a country whose shape combines a blend of compact and elongated features. Much of the territory is relatively compact but there is at least one elongated section (the protrusion). There is always the potential of isolation and disconnection between the protruded area and the rest of the state. In Southeast Asia, Myanmar and Thailand are classic examples of protruded states. Other global examples include Colombia in South America, Namibia in Africa, and Italy in Europe.

A **fragmented state** is a country comprised of parts that are physically separated, such as a group of islands or a mainland with separated sections apart from the mainland. Fragmented states have lots of potential problems in effectively controlling and communicating with the separated parts. In

Southeast Asia, The Philippines and Indonesia are classic examples of island countries while Malaysia is an example of a country where part of the state is on the Malay Peninsula and the other part is on the island of Borneo. Other global examples include the U.S., Argentina and Chile in South America, Denmark and Italy and Greece in Europe, and Japan in Asia.

A **perforated state** is where you have a country that completely surrounds another country, as if there is a perforation or opening in the territory. In Southeast Asia, Indonesia is now a perforated state, with the eastern part of the island of Timor (East Timor) recently becoming an independent country. Other global examples include Italy in Europe and South Africa in Africa.

Subregions of Southeast Asia

Southeast Asia includes the Mainland Subregion and the Insular Subregion.

Mainland Subregion

Mainland Southeast Asia is comprised of five states: Myanmar, Thailand, Laos, Cambodia, and Vietnam. These countries are all located on the southeastern mainland of Asia. Historically, this subregion was known as Indochina due to the extensive influences from China and South Asia. Buddhism is the dominant religion. There are varied multicultural and multiethnic groups that affect the Mainland. It is also a relatively sparsely populated area of Asia, with low levels of urbanization.

Myanmar is a protruded state that lies between India on the west and Thailand on the east. It is one of the world's poorest economies and was recently controlled by a repressive and abusive administration. Recently, the government has begun to incorporate some degree of press freedom and released a large number of political prisoners. The Irrawaddy River flows through the middle of the country and provides a rice-rich lowland basin in the southern part of Myanmar.

Thailand is also a protruded state. Its capital, Bangkok, is the largest urban center in the Mainland Subregion. Thailand occupies a central position in the Mainland Subregion and has recently experienced much economic progress. An indicator of its relatively higher level of development is that Thailand has one of the slowest rates of population growth in all of Southeast Asia.

Laos is a landlocked state with a relatively small population. Its economic infrastructure is poorly developed and most of the land, being mountainous, is not suitable for agriculture. Laos is primarily a Communist state and is increasingly pulled into the Chinese sphere of influence.

Cambodia is a compact state that is overwhelmingly Khmer in ethnicity and Buddhist in religion. When the U.S. was forced to withdraw from Southeast Asia in 1975, a radical group of young Communists, the **Khmer Rouge**, undertook a campaign of ethnic cleansing. As many as 2 million Cambodians were eliminated in what became known as the **"killing fields."**

Vietnam, along with Laos and Cambodia, was part of **"French Indochina."** After the end of World War II, Communist forces from China advanced southward into Southeast Asia, especially northern Vietnam. France, at the time, still controlled southern Vietnam. War broke out between the French and the Communists. The Geneva Agreement in 1954 partitioned the country into a Communist-dominated North and a non-Communist South. In the mid-1960s, the U.S. intervened in South Vietnam, Cambodia, and Laos to try to stop the whole area from turning Communist. The basis for American involvement was the **domino theory**. The domino theory stated the belief that if South Vietnam fell to the Communists, then Laos, Cambodia, Thailand, and Burma would fall, like the toppling of dominos. By the mid-1970s, the U.S. withdrew and eventually Vietnam was reunited into a single country, with Hanoi, the former Communist capital, as the capital of reunited Vietnam. Vietnam now has an emerging economy with increasing trade ties to the West, including the U.S.

Insular Subregion

Insular Southeast Asia is a physically fragmented area with Islam as the dominant religion. Like the Mainland Subregion, this is a multicultural and multiethnic territory. Insular Southeast Asia includes the countries of Malaysia, Singapore, Indonesia, and The Philippines.

Malaysia is a fragmented state with West Malaysia occupying the southern part of the mainland Malay Peninsula and East Malaysia occupying the northwest part of the island of Borneo. Most of the population and area are on the Malay Peninsula, including the capital of Kuala Lumpor. The majority of the populace is ethnically Malay and religiously Muslim.

Singapore is a city-state, with a modern infrastructure, no squatter settlements, and a high level of economic development. Singapore is one of the four **Asian Tigers** (South Korea, Taiwan, and Hong Kong being the other three) that rapidly grew from a less developed to a more developed economy. The country consists of a main island, just south of the Malay Peninsula. Given its relative location, Singapore has become one of the busiest ports and container terminals in the world. It is a classic example of an entrepot. An **entrepot** is a place that lies at the intersection of a number of trade paths, making it a major transshipment location. Goods are brought into Singapore, stored, and then reshipped to other places. This makes Singapore a **gateway** to and from Southeast Asia. Singapore is also a **development state**. The top priority of Singapore's government is to create and improve economic development. The government highly encourages foreign investment by multinational firms from the U.S., Europe, and Japan.

Indonesia is a fragmented state of more than 17,000 islands. It is a former Dutch colony, formerly known as the Dutch East Indies. The dominant island is Java, the country's core, with over 150 million people and its capital city of Jakarta. Java is one of the most densely populated areas of the world. Indonesia is the most populated country in the region and is the world's fourth-largest country in population. It is also the world's most populated Muslim country.

The Philippines is a fragmented state of more than 7,000 islands. It was a Spanish colony for centuries. As a result of the Spanish-American War, The Philippines became an American colony from 1898–1946. The Philippines became an independent country in 1946. It is the only predominantly Christian country in Asia, with Roman Catholicism being the main Christian faith. The Philippines has mostly an agricultural economy. Islam dominates in many of the southern islands and religion has been the basis for a lot of internal conflict in the southern Philippines.

AUSTRALIA AND NEW ZEALAND

Where Are Australia and New Zealand?

Australia and New Zealand are the most remote parts of the more developed world, located where the Pacific Ocean meets the Indian Ocean.

Major Qualities of Australia and New Zealand

Australia and New Zealand differ physiographically and climatically. Australia has a vast dry lowland interior while New Zealand is more mountainous and has a more temperate climate.

The region has an isolated location relative to the rest of the more developed world, being on the remote periphery of the more developed states.

Both Australia and New Zealand have a peripheral pattern of population and urban settlement, with most urban populations highly clustered on the coasts.

Australia has a rapidly changing human geography, especially through newer sources of immigration.

Australia and New Zealand are highly integrated into the Pacific Rim economies of Asia, essentially as providers of primary raw materials.

Commonalities of Australia and New Zealand

Australia and New Zealand share many common features. Both countries are economically well-developed with very high levels of urbanization. Australia and New Zealand share an excellent quality of life overall. They are both remotely located with the accompanying challenges of distance and cost. Both countries exhibit mainly a British heritage, with each experiencing concerns and issues with an indigenous population. With small populations, they share small, internal markets. Economically, they mainly export primary materials and import finished products.

Differences between Australia and New Zealand

Topographically, Australia is more low-lying while New Zealand is more mountainous. Climatically, most of Australia is dry whereas New Zealand has a more temperate humid climate. Australia is continental in size with New Zealand being much smaller in area. Australia has about 25 million people;

New Zealand only about 5 million people. Australia exports wheat, beef, wool, and minerals. New Zealand exports wool and dairy products.

Australia and New Zealand Problems

Both states have small populations, small internal domestic markets, and remote external markets. They each have major issues with indigenous populations. They have unpredictable trade links with Asia.

Distinctive Economic Patterns

Australia and New Zealand are both highly developed economies. Typically, more developed states import many raw materials and export finished products. Australia and New Zealand have a reversed trade pattern. They export raw materials and import finished products.

Subregions of Australia and New Zealand

Australia and New Zealand are the two countries that comprise this region.

Australia

Australia is the dominant component of this region, both in terms of area size and population. Australia is a continent-scale country that rivals China, Canada, the United States, and Brazil. About 25 million people inhabit Australia, making it one of the smallest more developed economies.

Australia is a federal state, comprised of six states and two territories. The six states are Victoria, including Melbourne; New South Wales, including Sydney; Queensland, including Brisbane; South Australia, including Adelaide; Western Australia, including Perth; and Tasmania, including Hobart. The two territories are Northern Territory, including Darwin, and the Australian Capital Territory, including the federal capital city of Canberra.

Australia is a highly urban country, with over 80 percent of the population being urbanized. Most of the population lives in scattered urban centers, principally on the southeast coast. **Sydney** is the largest city, with Melbourne being the second-largest city.

Historically, the original occupants were the Aborigines. The **Aborigines** are darker-skinned people who entered Australia about 50,000 years ago from Southeast Asia. Today, they are mainly concentrated in the Outback. The **Outback** is the dry interior of Australia, largely comprising the Northern Territory. Out of Australia's total population, only about 600,000 are Aborigines. The Aborigines were greatly discriminated and were largely looked at as second-class citizens. Today, they tend to be relatively poor, with higher levels of poverty and unemployment. It wasn't until the 1990s that the Aborigines could claim title to traditional land. And it wasn't until a few years ago that the Prime Minister officially apologized for their historic maltreatment. There are continuing aboriginal land issues, especially in the Outback.

The first European settlers were actually British convicts who arrived in the late 1700s. British prisons had become very overcrowded; many excess prisoners used to be sent to the British colonies in North America. However, with the end of the American Revolutionary War, Britain had to look for alternate sites. Australia became the largest of these destinations. In the early 1800s, more traditional

British settlers and farmers arrived. In the 1900s, large numbers of Europeans from other parts of Europe began to immigrate. More recently, Asia has become a major new source area for Australia immigration.

An advantage of the vast area size of Australia is a rich array of mineral resources. Australia has gold, oil, natural gas, coal, iron ore, uranium, and bauxite ores.

New Zealand

New Zealand is much smaller in land area and population than Australia. There are two major islands that comprise New Zealand: the North Island and the South Island.

The **North Island** is mainly of volcanic origin, contains the largest city, Auckland, and includes the capital city of Wellington. It is more populated than the South Island.

The **South Island** has been highly affected by glacial activity and is more mountainous than the North Island.

New Zealand lies on the **Pacific Ring of Fire**, making it very prone to earthquake activity, especially on the South Island.

There are only about 5 million people in New Zealand. The original inhabitants are the Maori. The **Maori** are of Polynesian descent and arrived in New Zealand about 1,000 years ago. Today, they comprise about 700,000 people, giving New Zealand a higher share of indigenous peoples compared to Australia.

OCEANIA

Where Is Oceania?

Oceania is comprised of tens of thousands of islands scattered over the Pacific Ocean between the Arctic and Antarctic polar areas.

Major Qualities of Oceania

Oceania is the largest of all world regions in territorial size, although it has the smallest land area and the smallest population. There are about 10 million people living in Oceania.

It is a highly fragmented region, comprised of three sets of islands: **Melanesia, Micronesia, and Polynesia**. New Guinea is the most populated island, with about 80 percent of the region's population.

Oceania has been strongly affected by the United Nations Law of the Sea provisions concerning the rights of coastal states over adjacent waters.

Oceania's islands are often divided into volcanic high-islands versus coral-based low-islands. **Volcanic high-islands** are economically based on agriculture. They tend to have more population. **Coral-based low-islands** are mainly oriented to fishing and contain smaller populations.

United Nations Conferences on the Law of the Sea

Historically, the territorial sea referred to the ocean waters adjacent to a coastal state where all the rights of a coastal state would prevail. The high seas were the larger areas of the ocean where everything was open to all. Over time, many disputes occurred based on varying sizes of areas being claimed by coastal states. From the 1950s to the 1980s, the United Nations convened several conferences on the law of the sea. In 1982, a Convention on the Law of the Sea established international guidelines for control of the oceans.

A **12-mile territorial sea** was established. Within this adjacent zone, a coastal state could exercise **sovereignty**, total control over all affairs. An **Exclusive Economic Zone** was established from 12–200 miles off a coastal state. Within this zone, coastal states could exercise limited economic rights over fishing and minerals. Beyond the 200-mile limit lay the **high seas**, open to one and all. The guidelines do not affect landlocked states nor do they effectively deal with issues where states are closer than 200 miles apart. They do provide an international framework that standardizes areas of control to minimize future disputes.

Subregions of Oceania

Oceania includes three subregions: Melanesia, Micronesia, and Polynesia.

Melanesia

Melanesia is a group of islands that lie northeast of Australia. The name refers to the darker skin of most of the residents. Melanesia is the most populous subregion in Oceania. Melanesia includes Papua New Guinea, Solomon Islands, New Caledonia, and Fiji.

Papua New Guinea is the most populated state in Oceania, with about 7 million people. It became independent in the 1970s after a century of British and Australian control. Papua New Guinea comprises the eastern half of the island of New Guinea. Crude oil is now Papua New Guinea's leading export. There are also deposits of gold, copper, and silver that help provide a diversity of resources.

The **Solomon Islands** are east of Papua New Guinea. There are approximately 1,000 islands, most unoccupied, that comprise the Solomon Islands. About 500,000 people live in the Solomon Islands. Numerous languages and physical fragmentation provide many conflicts for this area.

New Caledonia is a French territory directly east of Australia. There are about 300,000 people in New Caledonia; about 35 percent of the population is of European ancestry, mainly French. Nickel dominates the export economy.

Fiji is on the far eastern margin of Melanesia. It achieved independence in the 1970s. There are approximately 1 million residents living on about 100 islands. Over 40 percent of the population is from South Asia, brought from India by the British to work on sugar plantations.

Micronesia

Micronesia consists of more than 2,000 small islands that lie north of Melanesia and east of the Philippines. Subsistence farming and fishing dominate the economy but most islands survive on foreign aid. Until the 1980s, Micronesia was largely a United States Trust Territory. Now, it is comprised of a series of independent states: Marshall Islands, Northern Mariana Islands, Palau, Guam, and Nauru.

The **Marshall Islands** are in a free association with the United States. This is where the United States tested nuclear weapons in the mid-twentieth century.

The **Northern Mariana Islands** are a commonwealth in political union with the United States. Billions of dollars are sent by the United States in assistance to these countries.

Palau is a series of islands east of the Philippines that became independent in the 1990s. It is dependent upon the United States for financial aid and military security.

Guam is a United States territory where U.S. military installations and tourism provide most of the income.

Nauru got wealthy by exporting phosphate to Australia and New Zealand for use as fertilizer. Most of the main deposits have run out, providing much economic uncertainty.

Polynesia

Polynesia lies to the east of Melanesia and Micronesia, forming a large triangle stretching from the Hawaiian Islands to Easter Island to New Zealand. These many islands vary from large volcanic islands to low coral atolls. The Polynesians have lighter skin and wavier hair than most of the other peoples of Oceania. Polynesia includes the islands of Tonga, Tuvalu, Tahiti, and Samoa.

Tonga is a kingdom that became independent from the British in 1970. It lies southeast of Fiji.

Tuvalu also got its independence from the British in the 1970s. It lies east of the Solomon Islands.

Tahiti is a French territory, part of the broader French Polynesia that also includes the Marquesas Islands and the Society Islands. Tourism is a mainstay of Tahiti's economy.

Samoa is located to the northeast of Fiji and Tonga. The eastern part of Samoa has become very Americanized in society and landscape.

Appendix A

EXAM STUDY QUESTIONS

The Exam Study Questions refer to relevant pages in the Text, *World Regional Geography: An Introduction,* and an Atlas, *National Geographic Student World Atlas,* 4th edition.

Appendix B
Homework Assignments